IoT, Machine Learning and Blockchain Technologies for Renewable Energy and Modern Hybrid Power Systems

RIVER PUBLISHERS SERIES IN INFORMATION SCIENCE AND TECHNOLOGY

Series Editors:

K.C. CHEN, *National Taiwan University, Taipei, Taiwan*
and
University of South Florida, USA

SANDEEP SHUKLA, *Virginia Tech, USA*
and
Indian Institute of Technology Kanpur, India

The "River Publishers Series in Computing and Information Science and Technology" covers research which ushers the 21st Century into an Internet and multimedia era. Networking suggests transportation of such multimedia contents among nodes in communication and/or computer networks, to facilitate the ultimate Internet.

Theory, technologies, protocols and standards, applications/services, practice and implementation of wired/wireless

The "River Publishers Series in Computing and Information Science and Technology" covers research which ushers the 21st Century into an Internet and multimedia era. Networking suggests transportation of such multimedia contents among nodes in communication and/or computer networks, to facilitate the ultimate Internet.

Theory, technologies, protocols and standards, applications/services, practice and implementation of wired/wireless networking are all within the scope of this series. Based on network and communication science, we further extend the scope for 21st Century life through the knowledge in machine learning, embedded systems, cognitive science, pattern recognition, quantum/biological/molecular computation and information processing, user behaviors and interface, and applications across healthcare and society. Books published in the series include research monographs, edited volumes, handbooks and textbooks. The books provide professionals, researchers, educators, and advanced students in the field with an invaluable insight into the latest research and developments.

Topics included in the series are as follows:-

- Artificial intelligence
- Cognitive Science and Brian Science
- Communication/Computer Networking Technologies and Applications
- Computation and Information Processing
- Computer Architectures
- Computer networks
- Computer Science
- Embedded Systems
- Evolutionary computation
- Information Modelling
- Information Theory
- Machine Intelligence
- Neural computing and machine learning
- Parallel and Distributed Systems
- Programming Languages
- Reconfigurable Computing
- Research Informatics
- Soft computing techniques
- Software Development
- Software Engineering
- Software Maintenance

For a list of other books in this series, visit www.riverpublishers.com

IoT, Machine Learning and Blockchain Technologies for Renewable Energy and Modern Hybrid Power Systems

Editors

C. Sharmeela
Anna University, India

P. Sanjeevikumar
Aarhus University, Denmark

P. Sivaraman
Vestas Technology R&D Chennai Pvt. Ltd, India

Meera Joseph
Independent Institute of Education, South Africa

River Publishers

Routledge
Taylor & Francis Group

NEW YORK AND LONDON

Published 2023 by River Publishers
River Publishers
Alsbjergvej 10, 9260 Gistrup, Denmark
www.riverpublishers.com

Distributed exclusively by Routledge
605 Third Avenue, New York, NY 10017, USA
4 Park Square, Milton Park, Abingdon, Oxon OX14 4RN

IoT, Machine Learning and Blockchain Technologies for Renewable Energy and Modern Hybrid Power Systems / by C. Sharmeela, P. Sanjeevikumar, P. Sivaraman, Meera Joseph.

Routledge is an imprint of the Taylor & Francis Group, an informa business

ISBN 978-87-7022-724-7 (print)
ISBN 978-10-0082-440-7 (online)
ISBN 978-10-0336-078-0 (ebook master)

While every effort is made to provide dependable information, the publisher, authors, and editors cannot be held responsible for any errors or omissions.

Contents

2 IoT and its Requirements for Renewable Energy Resources 29
D. Gunapriya, R. Sivakumar, and K. Sabareeshwaran

8   Application of Machine Learning Techniques in Modern Hybrid Power Systems – A Case Study   173

B. Koti Reddy, Krishna Sandeep Ayyagari,
Raveendra Reddy Medam, and Mohemmed Alhaider

9   Establishing a Realistic Shunt Capacitor Bank with a Power System using PSO/ACCS   205

Ali Mohamed Eltamaly, Osama El Sayed Morsy,
Amer Nasr Abd Elghaffar, Yehia Sayed Mohamed,
and Abou-Hashema Ahmed

Preface

Renewable energy resources are alternative to fossil fuels and it always stands for '*new findings*' with *challenges* by the researcher to fulfill the power and energy demand. The book "IoT, Machine Learning and Blockchain Technologies for Renewable Energy and Modern Hybrid Power Systems" will provide an enhanced solution for various aspects of IoT, machine learning, and blockchain applications from the editors and diverse authors.

This book covers the different sections dealing with fundamentals and applications of IoT, machine learning, and blockchain technologies in renewable energy and hybrid power systems. It includes case studies like power quality monitoring for low voltage distribution systems through IoT, health monitoring of distribution transformers through IoT, blockchain with SHA-256, 384, and 512 application to renewable energy resources, etc.

It well describes topics with theoretical-based analysis and followed by numerical solutions and simulation results, case studies which make additional credit to readers for their future research or profession.

The chapters are lucidly covering the significant and bottle-neck challenges prevailing in the renewable energy and hybrid power systems, enabling the reader to better understand. The book will be readily available as reference materials for IoT, machine learning, and blockchain technology applications to renewable energy & hybrid power systems, and enabling the student community to create more interest and attention to take up the challenging renewable energy profession for their endeavors.

It is a unified contribution by international authors from Europe, India, China, Nepal, the USA, Egypt, Saudi Arabia, and Thailand.

Editors

Acknowledgement

Foremost, thanks to the Almighty for his everlasting love throughout this endeavor.

Acknowledgments are always a phrase to appreciate the resources and timely solutions either with the digital platform or real-time medium, timely support and a bond of encouragement is the vital tool for teachers and researchers from their Institutions. In these regards, we editors express our sincere thanks to Mr. S. Muthukumaran, Director, TECH Engineering Services, Chennai, India; Mr. S. Rajkumar, Executive, JLL, Bengaluru, India; Mr. K. Sasikumar, Electrical Engineer, Mott MacDonald, Noida, India; Center for Bioenergy and Green Engineering, Department of Energy Technology, Aalborg University, Esbjerg, Denmark, Department of Electrical and Electronics Engineering, KPR Institute of Engineering and Technology, Tamilnadu, India, Department of Electrical and Electronics Engineering, College of Engineering Guindy, Anna University, Chennai, India, Independent Institute of Education, Johannesburg, South Africa. Editors we got the full support and executed the task promptly where our Institution devoted the time and liberty for enhancement with research in particular to make this book a great success.

I wish One and All for the devoted time frame effort for the grand success of the book.

Editors

List of Figures

List of Tables

List of Contributors

Ahmed, Abou-Hashema, *Electrical Engineering Department. Minia University, Egypts*

Alhaider, Mohemmed, *College of Engineering at Wadi Addawaser, Prince Sattam bin Abdulaziz University, Saudi Arabia; E-mail: malhider3@gmail.com*

Anbarasi, S., *P S R Engineering College, India*

Ansari, Saniya M., *E & TC Department, Dr D Y Patil School of Engineering (DYPSOE), India; E-mail: saniya.ansari@dypic.in*

Ayyagari, Krishna Sandeep, *Department of Electrical & Computer Engineering, The University of Texas at San Antonio, USA; E-mail: krishnasandeep.ayyagari@my.utsa.edu*

Balaji, S., *IIT Kanpur, India*

Balasubramanian, T., *P S R Engineering College, India*

Calay, Rajnish Kaur, *Department of Building, Energy and Material Technology, UiT The Arctic University of Norway, Norway; E-mail: rajnish.k.calay@uit.no*

Chen, Xiaofeng, *Hangzhou Qulian Technology Co., Ltd., China; E-mail: chenxiaofeng@hyperchain.cn*

Elango, S., *Coimbatore Institute of Technology, India*

Elghaffar, Amer Nasr Abd, *CAlfanar Engineering Service, Alfanar Company, Saudi Arabia, Electrical Engineering Department. Minia University, Egypt; E-mail: amernasr70@yahoo.com*

El Sayed Morsy, Osama, *Alfanar Engineering Service, Alfanar Company, Saudi Arabia*

Eltamaly, Ali Mohamed, *Electrical Engineering Department, Mansoura University, Egypt; Sustainable Energy Technologies Center, King Saud*

University, Saudi Arabia, K.A.CARE Energy Research and Innovation Center, Saudi Arabia

Ezhilarasi, G., *Sri Sairam Institute of Technology, India*

Gunapriya, D., *Sri Krishna College of Engineering and Technology, India;* E-mail: gunapriyadevarajan@gmail.com

Jia, Xiangjuan, *Hangzhou Qulian Technology Co., Ltd., China;* E-mail: jiaxiangjuan@hyperchain.cn

Khakurel, Saju, *Electronics and Communication Engineering, Nepal;* E-mail: sajukhakurel9@gmail.com

Lakshmi, D., *Academy of Maritime Education and Training (AMET), India;* E-mail: lakshmiee@gmail.com

Maharjan, Asim, *Electronics and Communication Engineering, Nepal;* E-mail: maharjan291@gmail.com

Medam, Raveendra Reddy, *Department of EEE, Maturi Venkata Subba Rao Engineering (MVSR) College, India;* E-mail: raveendra_eee@mvsrec.edu.in

Mohamed, Yehia Sayed, *Electrical Engineering Department. Minia University, Egypt*

Mustafa, Mohamad Y., *Department of Building, Energy and Material Technology, UiT The Arctic University of Norway, Norway;* E-mail: mohamad.y.mustafa@uit.no

Ongsakul, Weerakorn, *Department of Energy, Environment, and Climate, School of Environment, Resources and Development, Asian Institute of Technology, Thailand*

Patil, Ravindra R., *PhD Scholar, Department of Building, Energy and Material Technology, UiT The Arctic University of Norway, Norway;* E-mail: ravindra.r.patil@uit.no

Punitha, K., *P S R Engineering College, India;* E-mail: kgpunitha@gmail.com

Ravi, C.N., *Vidya Jyothi Institute of Technology, India*

Reddy, B. Koti, *Department of Atomic Energy, India;* E-mail: kotireddyb@ieee.org

Sabareeshwaran, K., *Karpagam Institute of Technology, India;* E-mail: sabareeshwarank@gmail.com

Sanjeevikumar, P., *Aarhus University, Denmark;*
E-mail: sanjeevi_12@yahoo.co.in

Sharmeela, C., *Anna University, India; E-mail: sharmeela20@yahoo.com*

Sivakumar, R., *Akshaya College of Engineering and Technology, India;*
E-mail: Sivakkumar14@gmail.com

Sivaraman, P., *Vestas Technology R&D Chennai Pvt Ltd, India;*
E-mail: sivaramanp@ieee.org

Thu, Kaung Si, *Department of Energy, Environment, and Climate, School*
of Environment, Resources and Development, Asian Institute of Technology,
Thailand; E-mail: a.kaungsithu@outlook.com

Tiwari, Shubham, *Department of Energy, Environment, and Climate, School*
of Environment, Resources and Development, Asian Institute of Technology,
Thailand

Zahira, R., *BSA Crescent Institute of Science and Technology, India;*
E-mail: zahirajaved@gmail.com

List of Abbreviations

ACCS	Automatic capacitor control scheme
AI	Artificial intelligence
CPS	Cyber-physical systems
DEM	De-regulated electricity markets
DFIG	Doubly fed induction generator
DG	Distributed generation
DN	Distribution network
ECDSA	Elliptic curve digital signature algorithm
EMM	Energy management models
ESS	Energy storage systems
ETAP	Electrical transient analyzer program
FACTS	Flexible AC transmission system
FFNN	Feed forward neural network
FPA	Flower pollination algorithm
GA	Genetic algorithm
GPRS	General packet radio service
GPS	Global positioning system
GSM	Global system for mobile communication
IED	Intelligent electronic devices
IoT	Internet of things
IR	Infrared sensors
IRENA	International renewable energy agency
ITU	International telecommunication union
KNN	K-nearest neighbours
LCD	Liquid crystal display
LDR	Light-dependent resistors
LFC	Load frequency control
LSTM	Long-short term memory
M2M	Machine to machine
MG	Micro grids
MHPS	Modern hybrid power system
ML	Machine learning

MPPT	Maximum power point tracking
OSI	Open systems interconnection model
PI	Proportional integral
PID	Proportional integral derivative
PoW	Proof-of-work
PQ	Power quality
PSO	Particle swarm optimization
PV	Photovoltaic
PV	Predictor variable
RNN	Recurrent neural network
RTOS	Real-time operating system
RV	Response
SA	Smart appliances
SARSA	State action reward state action
SG	Smart grids
SHA	Secure hash algorithm
SLD	Single line diagram
VRES	Variable renewable energy resources
WRS	Wireless sensor network

1

Introduction to IoT

Asim Maharjan and Saju Khakurel

Electronics and Communication Engineering, Nepal
E-mail: maharjan291@gmail.com; sajukhakurel9@gmail.com

Abstract

This chapter discusses the field of the Internet of Things, or better known as IoT, and various concepts related to it along with its vast applications and various challenges faced by IoT in the present world. A bit of the history of IoT is also given so that the readers can get a sense of how long ago the concept of IoT was conceived before becoming the revolution it is today. It goes over some of the important milestones in the history of IoT from its naming to its widespread adoption. Some of the applications of IoT, namely, in households, healthcare, industries, and renewable energies are also discussed to show the fields that have already begun to integrate IoT technologies in their operation. A brief description of the enabling technologies of IoT is also given with the purpose that the readers will be able to grasp the overall working of an IoT system. It then explores the recent developments and achievements in the field of IoT to give readers a glimpse of the present-day landscape of IoT. Finally, the chapter also touches upon the many different kinds of issues and challenges that present-day IoT systems are facing. These range from the technical aspects of IoT such as compatibility, interoperability, and security to the more ethical aspects like privacy.

Keywords: Internet of Things (IoT), sensor, ubiquitous computing, home automation, Industry 4.0.

1.1 Introduction

With the exponential evolution of technology in almost every field like communication technology, digital technology, machinery, robotics, power and energy, and many more, it is common to have a few of these advancing technologies combined to form even more complex and advanced techniques and systems. In recent years, there have been numerous researches for the expansions in the existing systems to improve their efficiency and application in various sectors. It was not until a few years back when there were cable line telephones for communication. But due to the advancement and exploration in the communication field along with the accelerating development in wireless technology and digital technology, today, almost every means of communication is wireless along with the digitized systems embedded in it. And now with the rapid replacement of smart technologies among all the existing systems, almost all electronic gadgets are as smart as a human task would be.

Following this, the idea of convergence of multiple technologies gives rise to various newer ideas for the implementation of the existing technologies for a better and dynamic system. Internet of Things (IoT) is one of the mixtures of such multiple existing technologies, which incorporates multiple technical fields such as sensors, the Internet, software, real-time operating system (RTOS), embedded systems, etc. The idea of IoT is to act as a bridge that connects physical things from the real world to the virtual world through the Internet. This concept of connecting the physical and the virtual worlds was first initially proposed by Mark Weiser in the early 1900s, where IoT allows physical objects of the real world to be able to be remotely controlled via the advancing technologies through the Internet [7]. Later in 2005, the idea of the IoT was officially introduced by the International Telecommunication Union (ITU) at the World Summit that was held at the Information Society in Tunisia. They also released an ITU Internet report which included an in-depth knowledge about the IoT, its concepts, and its global effect around the globe [24].

IoT is a large networking platform for interaction and control of multiple electronic tools via the Internet. In simple words, it is a combination of sensors, processors, and then controllers for actuation, which all communicate through the use of the Internet. The data streams, from the sensors or any other smart devices that acknowledge the change in environmental factors, get stored up in a common platform where the necessary information gets processed, computed, and analyzed. This filtered data can either be stored for future reference or can be used immediately for taking actions like controlling

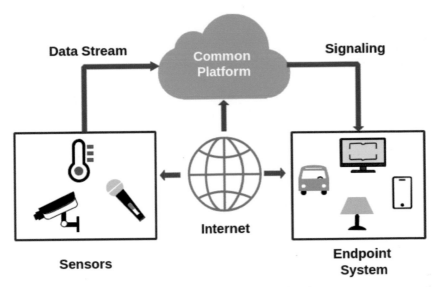

Figure 1.1 Related concepts of IoT.

or switching the other devices sharing the common platform in the network. It carries the idea of independence of human interventions in monitoring and control of any digital systems for efficient work through. These systems do not possess any delay, minimize human effort by expanding the independence of humans, efficiently utilize the resources, save time, and are transparent. Figure 1.1 shows the concepts of IoT.

With the expanding knowledge, the unlimited possibility of implementation areas, and exponential advancement in technology, almost every human task is replaceable with a monitoring and control system. Few sensors for observing the changes in respective parameters of a dynamic environment, a processor for implementing the required algorithm, and controllers or actuators for maintaining the desired state. This idea expands the horizon of IoT which now, with few advanced types of equipment, can be implemented in almost every sector; for example, self-driving cars, medical recordings, smart home systems, manufacturing to customer care, irrigation systems, educational institutions, e-commerce, and so on. Figure 1.2 shows the various elements that are involved in the monitoring and control of the physical system through IoT.

A smart home can be considered the best example of IoT. The smart home system incorporates all the features that an IoT has to offer along with providing a clear concept on "What exactly is an IoT?" As mentioned

Figure 1.2 Various elements involved in the monitoring and control of physical systems through IoT [25].

earlier, IoT is a system where the physical objects are connected to a common virtual platform for interacting with every other device in the network, which are programmed to perform certain tasks that require less to no human interventions.

In a smart home system, multiple sensors are embedded in different locations of the house that tracks down the changes in a state like temperature, lighting, air quality, etc. These sensors are the starting point in the system whose data are then collected to a virtual cloud via the Internet. The data in the cloud is then processed and filtered after which the processed data is analyzed in real time. The endpoint to this system can be the APIs or the controllers in the home itself. This system allows the user to make changes in their homes when no one is physically present there or just monitor the activity that is taking place at the home when they are away. Either way, this system excludes the human intervention, establishes a virtual connection among the multiple devices, and takes necessary actions as it is programmed to.

Along with the existing system, there are numerous possibilities to increase its ability of performance such as a system that connects the user phone to the home system, calculates its distance from the home and unlocks the doors, sets the temperature, and turns on the lights on the arrival of the user. Furthermore, the expandability of these systems is just limited by human ideas, and with the massive developments that have been occurring in this field, there soon arrives a future where just the thought of outdoors triggered in the brain can cause the car to be ready at the front door.

1.2 History

The term Internet of Things was first coined by Kevin Aston in 1999, in a presentation for Procter & Gamble linking the concepts of radio-frequency identification (RFID) to the company's supply chain [1].

However, the first use of the IoT, as we know it now, was even before the term was coined. The concept first appeared in the mid-1970s at the Computer Science Department of Carnegie Mellon University in a Coke vending machine. It allowed students to view the status of the vending machine, such as knowing when the vending machine was full/empty or whether the drinks inside the vending machine were hot or cold. The vending machines had micro switches that were used to detect the coke bottles. These switches were then connected to the university's mainframe computer based on PDP-10. The students could then inquire about the status of the vending machine through an inquiry program written for it.

Back then, the Internet as we know it today did not exist. The communications were mostly carried out through ARPANET, which would later become known as the Internet [2].

This is just one of the examples of the concepts of IoT being conceived before the term was even coined. Many authors have written about devices interacting with each other through wireless means and have talked about such concepts in great detail. It is only now that such concepts have taken concrete forms.

As the years progressed from the mid-1970s, the Internet became more available to the general public. With large businesses adopting the Internet, the vast possibilities tied with the Internet rapidly started coming to light. With computers connected to every part of the world, sharing information was trivial. Along with the development of the Internet, computers were becoming smaller, faster, and more available. In 1991, author Mark Weiser wrote the article "The Computer for the 21st Century" [7]. The concept of Ubiquitous Computing began with it. The author depicted many real-life situations illustrating the widespread adoption of computers in performing day-to-day tasks such as making coffee, reading newspapers, identifying oneself, and so on. The author also noted how interconnections among these devices would present many challenges to the networking infrastructure back then. In the present, however, network infrastructures are very developed and network bandwidths of up to Gigabits are readily available. Thus, the future depicted by Mark Weise has certainly turned out to be a reality in today's world.

Attempts were made in the 1990s to have devices connected by following a set of standard protocols. Microsoft's "At Work" and NEST were two big technologies that attempted to embody the spirit of IoT. However, neither of them saw any widespread use.

IoT would see its emergence during 2008-2009 [3]. The first international conference of IoT was held in Zurin in 2008, where 23 out of 92 submissions were accepted [4]. Following the first international conference, many other such conferences related to the field of IoT were also held. Thus, IoT started gaining rapid popularity among scholars and tech enthusiasts alike.

Today, IoT is one the most rapidly growing fields in technology with vast applications in business, economics, health care, industry, power and energy, electric vehicles, and so on. With everything being smart and connected to the Internet, the need to connect them to achieve accessibility of orders of magnitude greater grows stronger every day.

1.3 Applications of IoT

1.3.1 Domestic Applications

IoT has vast applications in domestic fields. Applications range from the usual household monitoring to waste management and home security.

Household monitoring is one of the most known examples of IoT in households. With the help of IoT, people can monitor various equipment in their homes, such as lights and other electronic appliances. With advancements in technology and several everyday household appliances like refrigerators, television, etc., becoming smart, they can be connected to the Internet and can provide us with valuable data about their usage. This data can be accessed from anywhere in the world through the Internet. Another one of the major applications is switching, i.e., the act of powering ON/OFF devices through the internet. With this, we can effectively control our house from anywhere in the world.

Besides this, IoT can provide us with sophisticated data such as power usage, temperatures, amount of oxygen, and so on. Analyzing these data and performing the necessary changes to optimize these variables can help people dramatically improve their living conditions. Power utilization around the house can be made more efficient by powering ON/OFF certain appliances based on the data collected through sensors, e.g., light in the room automatically turns OFF when there is no one in the room.

One of the famous examples of such devices for home automation based on IoT is the Amazon Echo, which allows users to control many other devices such as speakers, televisions, and so on [23]. It even has additional features such as voice commands, web search, temperature sensing, etc., which provides immense accessibility to the average person. Thus, the power of IoT in the household department can be readily seen from this example.

A distinct feature of household automation is centralized control, i.e., controlling many appliances using a single device such as your mobile phone, computer, etc. As an example, web interfaces can be built, through which you can change the temperature of your smart heating system, control the room temperature of your smart AC, turn OFF the light, view your electricity usage, and many more.

With such vast applications in the domestic field and the ease of availability of hardware devices, many enthusiasts are also attracted to building such systems on their own. Thus, domestic applications remain one of the most prominent areas for deploying IoT-based solutions.

1.3.2 Applications in Healthcare

IoT has a wide range of applications in the health industry. The term Internet of Medical Things (IoMT) was introduced to specifically refer to the application of IoT in healthcare. In an infographic published by Deloitte, over 60% of EU health organizations have increased their adoption of digital technologies and have also adopted virtual ways of interacting with patients [10].

The adoption of digital technologies provides doctors and other medical professionals with new avenues for medical diagnosis and patient monitoring. With IoT-enabled devices, remote health monitoring and emergency notification systems can be put to use [11]. Many such devices have been developed and are in a constant state of improvement. Devices for heart rate monitoring are present today in the form of smartwatches and are capable of providing sophisticated levels of information such as heart rate, irregular heartbeat notifications, ECG reports, and even sleep monitoring. The devices can also be used to provide medical information, such as the dosage and when to take it, and alerts as well as for monitoring whether a patient has taken the medicine or not. Through the use of IoT, the data generated by such devices can be provided to doctors and physicians who can analyze them and provide better medical diagnoses to the patients. It also allows people to monitor themselves their own health conditions and improve their lifestyles to achieve better health.

Another promising application of IoT in the healthcare industry is in patient record management. All the medical records of patients can be stored and then made to be readily available to the patient and other doctors and professionals. This way, doctors who are treating the patient will have access to all the prior records which can lead to a more efficient and accurate diagnosis [8].

In addition to IoT, machine learning and data analytics have also seen widespread adoption in healthcare. Machine learning in combination with IoT can provide a very powerful tool for medical diagnosis. In [12], the authors have proposed a predictive machine learning algorithm for the detection of heart diseases by using data such as blood pressure, heart rate, respiratory rate, and blood sugar. In [13], an IoT system for monitoring the health condition of patients with high blood pressure is proposed by using HRV parameters. Many similar studies regarding the field of data analytics in conjunction with IoT for the medical sector can be found in [14].

1.3.3 Applications in E-commerce

IoT has also found its use in the E-commerce sector. One of the applications is in supply chain management which deals with handling the flow of goods and services from production to delivery to the customers. Rapid digitization has also affected this area very much. With the help of IoT, the information of each and every item in the warehouses can be made available, and product identification and tracking in each stage in supply chain management can be achieved using RFID [26]. Data such as temperature, humidity, etc., are of much importance when the storage of goods is concerned. Video cameras have been used for surveillance for a very long time throughout history for security purposes. Similarly, automated unmanned vehicles also provide a great productivity boost in the supply chain as this greatly reduces the labor cost. Consequently, the entire warehouse can be monitored effectively [27]. The survey in [15] highlights the value of IoT-based systems in SCM. In that survey, 12 participants from various companies belonging to a higher rank in SCM were interviewed about the effects of adopting an IoT-based approach. All the participants were very positive about the effects of IoT in their day-to-day work. Thus, IoT can play a big role in supply chain management.

Besides this, IoT can also greatly improve the consumer experience in the retail market. With online shopping/marketing becoming the norm nowadays, this field has already aggregated IoT in its business model. Information about the items in a retail shop is readily available through various outlets such as websites, mobile applications, and such. Food menus in restaurants along with the vacancy and reservations are only a few clicks away. Many applications are now available that allow people to compare prices between various goods and services provided by different providers which can greatly aid in the decision-making process for a consumer. Online payment systems are also gaining a lot of traction. With online payment services becoming more accessible to retail shops along with the convenience of being able to pay without using physical cash, online payments will likely become the preferred way of payment in the near future.

1.3.4 Industrial Applications

In the industrial sector, IoT is better known as the industrial Internet of Things (IIoT). IIoT represents what is known as Industry 4.0 [28], the new revolution in the industrial sector brought on by the use of IoT, data analytics, cyber-physical systems (CPS), cloud computing, edge computer, and other modern technologies. Many technologies related to IoT such as wireless sensor networks (WSNs) and RFID are heavily used for the identification and tracking of goods in industries. WSNs enable industries to accurately sense and monitor their environmental conditions by utilizing a network of sensors placed throughout the manufacturing plant [29]. It allows them to monitor the present conditions and make necessary changes to get maximal product output. Intelligent process control methods with the help of a wide variety of sensors can be applied in the traditional manufacturing process which can further boost productivity.

Besides manufacturing, IoT also provides a lot of benefits in the agricultural sector. A large number of sensors connected can provide farmers with data such as soil humidity, temperature, and pH levels of the entire farm. This allows the farmers to make proper adjustments to the ever-changing environment to get better crop yields. RFID is widely used in livestock farming for tracking animals, their behaviors, and their physical conditions [16]. The physical conditions of individual animals can be tracked easily and be made available to the farmers who can take better care of the animals. Sensors can provide information such as the location of cattle when they are grazing and their health condition such as temperature, heartbeat, etc. Any anomaly in their health conditions can now be easily detected and swift treatment can be done. Thus, productivity in agriculture and farming can be greatly increased by the use of IoT-enabled technologies.

1.3.5 Applications in Energy

IoT has found great applications and huge potential in the field of energy. As it has been discussed over several applications, one of the major uses of IoT is for monitoring processes. This works particularly well for increasing the efficiency of energy utilization.

Various energy monitoring systems have been proposed using IoT, such as those given in [30–32]. These systems monitor the power consumption and allow the users to visualize the energy usage of their households and buildings. This allows people to more efficiently utilize their electricity consumption by preventing unnecessary usage of electrical appliances. When released to the general public, small energy savings can greatly add up and lead to a lower energy consumption overall.

IoT has also seen several applications in the field of managing and monitoring solar energy. Solar power is greatly influenced by environmental factors such as the intensity of the sun, cloud movements, etc. [33]. Therefore, we require constant monitoring for the efficient and reliable operation of solar systems. A solar facility with IoT capabilities was developed in Arizona State University park [34, 35]. It allowed users to obtain various information related to the solar plant operation such as cloud movement and solar panel analytics. The smart monitoring devices installed were also able to perform more sophisticated tasks such as fault detection, bypassing of faulty panels, changing panel connections, etc. Many other solar panel monitoring systems have also been proposed such as those in [36–38]. These systems allow users to monitor the voltage, current, and temperature of the solar panels remotely. Similarly, IoT can also be used to monitor wind turbines. Variables such as wind speed, rotation of turbines, generated output, etc., can be monitored from the turbines. Since turbines are generally placed in remote windy areas, the wireless nature of IoT can provide further benefits in monitoring and control of the turbines. IoT can also be used for the automated control of solar as well as wind-powered plants, which can greatly help increase efficiency and power generation. By monitoring variables such as wind direction, cloud formations, etc., we can adjust the parameters of solar panels and wind turbines to get the maximum energy generation from them.

Another application of IoT in the energy field is in the monitoring of smart grids. The smart grid is a power grid that can deliver power in a controlled way according to the demand of the consumers [39]. These grids intelligently integrate the power consumption and supply according to the available as well as produced energy. Reliable and real-time information are very important in smart grids for robust and efficient delivery of power in smart grids [40]. To this extent, IoT provides a general framework to communicate information between devices in a smart grid. It can also be used for monitoring and metering the various power plants, domestic houses, and electric vehicles which are all a part of the distributed smart grid network. IoT in smart grids can also be used to monitor and collect consumer data which, when coupled with data analysis techniques such as machine learning, can be used to create energy management models (EMMs) [58]. The EMM can then be used to optimize the performance of the smart grid. They can also be used to predict and analyze consumer energy demand in the future. Therefore, IoT has a huge potential in the future in the field of energy.

1.4 Technical Details of IoT

1.4.1 Sensors

Generally speaking, sensors are those devices that convert physical phenomena like temperature, pressure, sounds, etc., into electric signals. Sensors provide us with the means to observe our world in terms of electrical signals. These signals are generally converted into a digital form and then processed.

For example, consider the temperature sensor measuring the temperature of a greenhouse. In this case, the greenhouse will act as a plant, i.e., a system which is being monitored by us. The temperature inside the greenhouse is about 30°C. The temperature sensor will give the output in the form of voltage, say around 4 Volts. This voltage of 4 Volts will then correspond to a temperature of 30°C. Furthermore, these 4 Volts will be converted into digital form, i.e., in terms of bits by using a special device known as analog-to-digital converter (ADC). If we assume that the ADC output is 1101 0000 bits or 208 in decimal, then this value of 208 will now correspond to the temperature of 30°C. In most cases, 208 will immediately convert into a numerical value of 30°C to make processing easier. This example only highlights the major components involved in extracting data from sensors, namely the *plant* (greenhouse in our case), the sensor itself, a device to convert the analog reading into a digital form, and finally a processor to act upon this acquired data as shown in Figure 1.3.

This is only a relatively simple and a small example of how a sensor integrates with the IoT environment [59]. The processing device will then act as an interface for the IoT system to interact with the environment that the

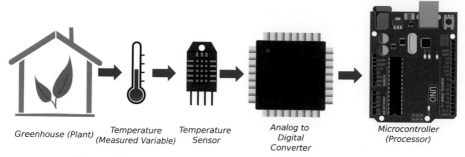

Greenhouse (Plant)   Temperature   Temperature   Analog to   Microcontroller
(Measured Variable)   Sensor   Digital   (Processor)
Converter

Figure 1.3   Example of an IoT system – temperature monitoring.

sensor will be monitoring. The complexity of the sensing system depends on a large number of factors such as the plant being monitored and the phenomena being monitored. Nowadays, many sensors even come with built-in ADC and thus only need some form of communication with the processing device which further simplifies the work done by the processor.

Some of the sensors used to monitor the physical devices and their description are given below.

A. Temperature Sensor

As the name suggests, these sensors measure the temperature of the environment. They do so by using thermistors, thermocouples, and other semiconductors. Figure 1.4 shows the typical sensor used to monitor both temperature as well as humidity.

B. Soil Humidity Sensor

As the name implies, soil humidity sensors are used to monitor soil humidity in agricultural land. The typical soil humidity sensor is shown in Figure 1.5.

Figure 1.4   A temperature plus humidity sensor.

Figure 1.5 A soil humidity sensor [51].

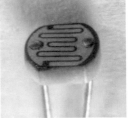

Figure 1.6 LDR (left) and photodiodes (right).

C. Photodetectors

The photodetectors are used to measure the intensity of light in the environment. Two common light sensors are the light-dependent resistors (LDRs) and the photodiode. An LDR varies its resistance according to the amount of light falling on it. Photodiodes, on the other hand, work on the principle of photoelectric effect where a photon from a light source striking a semiconductor produces electrons, which, in turn, causes current to flow. The typical LDR and photodiode are shown in Figure 1.6.

D. Infrared Sensors

Infrared (IR) sensors are used to detect IR radiations that are invisible to the human eye. They have a wide number of applications such as object and motion detection. IR can also be used for data transmission. These sensors can be active or passive. An active sensor has a built-in IR generation mechanism as well as receiving mechanism, whereas the passive IR sensor only detects the incoming IR radiations. The typical passive IR sensor is shown in Figure 1.7.

Figure 1.7 A passive IR (PIR) sensor [53].

Figure 1.8 An ultrasonic sensor [51].

E. Ultrasonic Sensors

Ultrasonic sensors are used for measuring distance. They work by measuring the time difference between the transmitted and the reflected ultrasonic waves which can then be converted into the distance by using the speed of the ultrasonic wave. A typical ultrasonic sensor is shown in Figure 1.8.

F. Gas Sensors

Gas sensors are used to detect the presence and concentration of various kinds of gases in the atmosphere. One of their major uses is in air pollution monitoring, where the sensor is used to measure the harmful gases present in the air such as CO_2, carbon monoxide, etc.

G. Atmospheric Pressure Sensor

Atmospheric pressure sensors are used to measure the atmospheric pressure and also sometimes as an altimeter for measuring the altitude of aircraft, spaceships, etc.

1.4.2 Actuators

Actuators are those components that convert an incoming signal such as an electrical voltage into a physical form such as a force, velocity, light, temperature, etc., or even into a non-mechanical form such as voltage and current itself [42]. They receive an input signal from a controller and produce an output that depends on the input signal. The output of the actuator interacts with its surroundings to bring some kind of change to it. As an example, a DC motor can be thought of as an actuator with an electrical voltage as its input signal and the rotation of its shaft as the output. The rotation of the shaft can interact with the physical world in several different ways, whether it is changing the position of the system by driving some wheels, opening or closing a door, pumping water, etc. Depending on the type of input signals and the corresponding output signal, actuators can be classified into several classes such as hydraulic, pneumatic, electric, mechanical, and so on [41].

In IoT-based systems, electric actuators are dominant primarily because they can be controlled using electrical signals. Some examples of electric actuators are DC motors, servo motors, electromechanical relays, etc. Electric actuators are also versatile and flexible in the sense that input signals can be controlled by the software running in the processor which allows for much better control over their output. For example, the speed of a DC motor and the position of a servo motor can be easily controlled through software. Thus, by including programming logic, a single electric actuator can be used for a vast number of applications. Furthermore, a processor can use the input data from the sensors to generate the input signals for actuators. This is a fundamental aspect of smart devices where devices can produce outputs according to the inputs without any human interference [42]. The typical example of the automatic door opening and closing is shown in Figure 1.9.

Figure 1.10 shows the simple automatic door opening system in which a DC motor is used as an actuator whose goal is to simply open and close the door. The camera acts as a sensor that gives pictures as input to the processor. The processor is then tasked with person detection. When it detects someone approaching the door, it gives an electrical signal to the DC motor to open the door. Obviously, in an IoT-based system, we would be able to do much

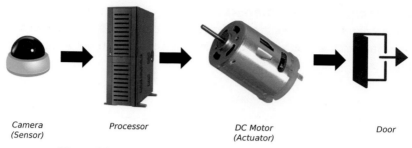

Camera
(Sensor)

Processor

DC Motor
(Actuator)

Door

Figure 1.9   Example of an automatic door opener [49], [50].

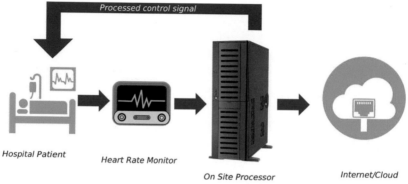

Hospital Patient

Heart Rate Monitor

On Site Processor

Internet/Cloud

Figure 1.10   Patient monitoring, an example of on-site processing [50].

more than just opening and closing of doors based on the camera image. With multiple of these systems placed at various locations along with popper person identification, we can track people in buildings solely based on which doors they have passed through.

1.4.3 Processing Topologies

Data processing is a very important part of IoT systems. With vast amounts of data collected from the physical system, it is of utmost importance to have processing capabilities that can work with huge amounts of data efficiently. Several things have to be considered for choosing the right processing topology. Data considerations are some of the most important factors. For example, IoT-based systems can be real time, where data is collected and processed in real time. These systems are generally process monitoring systems such as monitoring the temperature in a room, the pressure inside a chemical reaction chamber, monitoring the heart rate and blood pressure of a medical patient, etc. These systems involve a comparatively smaller workload; however, the latency must also be very small. Here, latency would mean the time between when the observation was taken (i.e., when the heart rate of a patient is measured) and when the processor would produce the corresponding output. On the other hand, IoT systems also have to deal with non-time critical data such as soil/humidity moisture from agricultural farms. These data can be stored in a database, possibly in a data center, and can be processed later on. Thus, latency is not a very important factor for these types of systems. The level of sophistication required is also very less compared to real-time systems in which every component in the IoT system must run with minimal latency for proper functioning [43].

Because of the nature of real-time systems, it has become necessary to add computing power near the source of data so that the latency is minimized as much as possible. Thus, two topologies are naturally born from this: *on-site processing* and *off-site processing* [17]. On-site processing refers to the data processing that occurs at the source itself. Here, the sensors are coupled with devices that have significant processing capabilities. These devices may be microprocessors, microcontrollers, or even large mainframe or supercomputers depending on the need. In this scenario, it is no longer dependent on the network for computing and, therefore, in some ways, more reliable for real-time systems. It has the added benefit of needing smaller bandwidth. The data collected can be shared in the IoT system at a later time. Here, the processing of data takes a higher priority than the data being available to the end-users. Figure 1.10 shows the typical on-site data processing system.

On a similar note, we have off-site processing. Here, the processing is done remotely, i.e., not at the source of the data. The data generated at the source has to be transmitted somehow to the processing site where it is processed and the processed results are then transferred to the users. Thus, it will generally have higher latency since it involves the transmission and reception of data over a network. It is generally used when the processing of data can be delayed with no repercussions. One of the main advantages of off-site processing is that it can greatly reduce implementation costs. Multiple sensor nodes can utilize a single processing unit compared to on-site processing where every sensor node needs to be paired with a processing unit. They can also be used for heavier workloads on the data since off-site processing gives us more flexibility in choosing more powerful hardware. As an example, consider a retail store that changes its goods based on customers' preferences. For this, the data of customers' preference of items are logged. Off-site processing is preferable for this since the store needs to collect a large

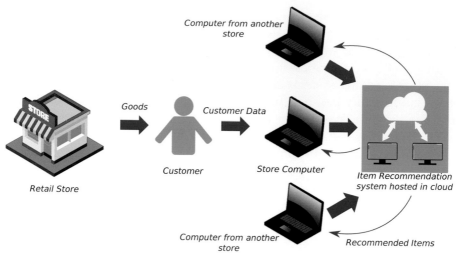

Figure 1.11   Retail stores, an example of an off-site processing topology [50].

amount of data from the customers before ordering new goods in the store. Also, the store can use pre-existing systems for this task. The data logged in a day can be sent to a cloud service, such as an item recommendation system, which can perform data analysis on the customer data to figure out the customer preferences and the kinds of items that customers are potentially looking for. The cloud service can also be made available to be used by other retail stores. This way, every store does not need to deploy its item recommendation system and can use pre-existing ones, thus saving a large amount of money and time. An illustration of an off-site processing system is shown in Figure 1.11 [17].

1.4.4 Communication Technologies

Communication forms the basis for devices to transfer and share their data in IoT. To this end, there must be a connection between the devices and certain protocols have to be followed. The open systems interconnection (OSI) model is a popular model on which many different communication systems are based [44]. The OSI model defined seven layers [45], with each layer having its unique functionality and also utilizing the functions of the layer as listed in Figure 1.12.

The seven layers combined provide the full communication service from one device to another. The physical and the datalink layers are responsible

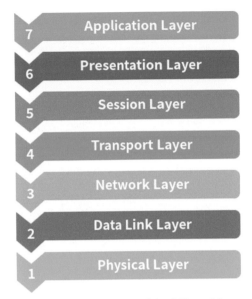

Figure 1.12 Layers of the OSI model.

for data transmission from one device to another through a single connection; for example, from the sensor to a microprocessor or your phone to the nearby Wi-Fi router. The network layer is responsible for data transmission through the network, i.e., from the sensor node to your computer. The transport layer is responsible for data transmission between processes, i.e., from the sensor node to the program that is using the sensor node. The above layers deal with data presentation and processing. These layers are generally present in every communication system in one way or another.

The bottom four layers of the OSI layer are of particular importance. Because of the wide variety of devices that are used, there is also a large variation in the technology used in each of the layers. The physical and the data link layers consist of various wired and wireless communication protocols used to connect two devices through a single link. Generally, sensors are connected to microprocessors through a wired channel. The sensor can communicate with the processor using various protocols such as I2C, SPI, USART, and so on. The processors, being more powerful devices, are then tasked with sending the data over the network which can be done through a wired channel such as Ethernet or a wireless channel. IEEE 802.15.4, Zigbee, LoRa, Bluetooth, Wi-Fi, and RFID are some of the popular technologies used for wireless communication [17, 46].

Table 1.1   Different types of the protocol used in IoT systems.

	Functionality	Example protocols
Infrastructure protocols	They handle the routing of data so that it reaches the correct destination. These are very dependent on the type of network used, e.g., Internet.	IPv4, IPv6, QUIC, nanoIP, etc.
Discovery protocols	They are used to discover whether a certain device or a certain service is available in a network or not. Besides this, they are also used when a new device is connected to a network to establish proper communication with the device.	mDNS, UPnP, etc.
Data protocols	They deal with data transfer, storage, and distribution. Data transmission can occur between one source and multiple destinations. Thus, these protocols have to be adept at handling large amounts of data transfer.	MQTT, REST, WebSockets, etc.
Identification protocols	They are used to identify devices in the network uniquely.	EPC, uCode, URI, etc.

The other layers then deal with the transmission of data from the source endpoint to the destination endpoint. Besides this, they also provide other important functions such as device discovery, identification, and management. Thus, we can classify the various protocols used in terms of their usage into the groups as listed in Table 1.1 [17].

1.5 Recent Developments

With the advancements in extensive digital technology and progressive optimal algorithms, IoT is emerging in almost every sector as possible and with its capacity expanded to virtual memories as well as the scalability factor. IoT is now on the verge of taking over the services by Internet and electronic gadgets to a newer level. For a smart future, IoT is the key source impacting the larger fields like big data and smart cities. Furthermore, due to its quick adaptability, increasing capability, better security, larger scalability, and reusability, it is safe to say that IoT is going to have a touch on everyone's day-to-day life activities. The whole idea of IoT stands on data entry, its manipulation, and necessary actions, but with the wide application area and increasing data levels including some confidential information too, the future of IoT leans toward complicated security and privacy issues. This concern is captivating the concentration from both industry and academia for performing research on IoT's security and privacy. Along with the security issues, the future of IoT also concerns the management issues like monitoring, updates, diagnostics, crash analysis and reporting, processor speed, power consumption, OS, standards and units, platforms, and so on. Hence, the recent development in IoT is concerned with these factors, especially prioritizing the security and privacy issues.

Since IoT connects multiple devices into a common network, the safest and secure point-to-point connection is preferred. Various existing standards have been used in attempts to solve the interconnection issues between the devices in the network. Also, various alliances and standards such as ITU-T, IEEE P2413, Allen Alliance, Open Connectivity Foundation, and IPSO Alliances have suggested a framework for IoT and connectivity of local IoT devices. OneM2M is one such proposed standard that aims to provide integration and interconnection between various smart IoT devices. The oneM2M project was founded by the Standards Developing Organizations (SDOs) of many countries such as Korea, USA, Japan, and so on [18].

The integrated circuits designed for IoT systems are focused on specific purposes and thus mainly utilize a robust and powerful system on chips (SoCs) for their operation. With security always being one of the most important aspects of an IoT-based system, IoT systems need to have security measures to protect the user data. Many different approaches for security in IoT have been proposed in recent years using new technologies like artificial intelligence (AI), machine learning, big data, and blockchain [55]. Some approaches using blockchain to improve security and privacy have been discussed in [53] and [54]. Similarly, a discussion on other approaches using machine learning can be found in [56] and [57]. Utilizing the SoC nature of IoT devices, a physically unclonable function (PUF) based cryptographic security solution for IoT SoCs has been proposed which deploys a mechanism using PUF and symmetric cryptography to provide a secure means for data communication [47].

Besides the advancement that is dedicated solely to IoT, advancements in other fields are also showing their impacts on bettering the IoT. The upcoming generation of mobile technology is predicted to highly benefit the IoT field. This is because of the invention of 5G technology though its specifications are yet to be finalized. With the greater speed of connectivity and lower latency, it will be aiding in speeding up IoT devices and help in better remote access. Another major emerging field is that of artificial intelligence (AI). Integration of AI with IoT can be a boon to the world of innovation. Using the data obtained from smart IoT devices in order to train themselves as smarter as a human brain can be a game-changer to the world of automation. Self-driving cars are a great example of this combination. Cloud computing is another major area that is immensely alternating with every other data storage and manipulation system available. The field of robotics has also seen the use of IoT and other cloud technologies. This is referred to as cloud robotics that aims to utilize IoT and other cloud services [48]. It allows robots to share information and utilize many powerful resources that are distributed in the cloud.

1.6 Challenges

Like many modern technologies, IoT also faces many challenges that prevent it from widespread adoption. IoT faces challenges at both the technological levels and at the ethical level. From a technological perspective, IoT faces challenges such as interoperability, scalability, security and privacy concerns, data concerns, etc.

With a large number of IoT systems being deployed, there is a higher chance of interoperability issues, i.e., integration of different systems into a single one. Many systems are proprietary and thus incompatible with each other. Making them compatible would require compatibility at all levels such as in hardware, protocols, etc. [19]. In order to overcome these issues, standards need to be developed which cover the technical requirements including hardware and software requirements for all the devices for better interoperability. Various efforts have been made by standards organizations like ITU, ISO, IEEE, IEC, etc., to build standards for fundamental IoT technologies like RFID, WSNs, etc. [20]. Scalability, on the other hand, brings along a lot of other major issues. Most IoT systems have to be designed and used on a large scale. This brings about the concerns of security and privacy among many others. IoT is an amalgamation of all kinds of technologies that raise security concerns at all levels. Each part of an IoT ecosystem must be secure against vulnerabilities and be robust against outer disturbances and interference. A lot of devices used in IoT have limited computational capabilities, which makes it very difficult to implement robust security measures. This can open up the entire ecosystem to vulnerabilities. One famous example of IoT security being compromised is the Mirai botnet [21]. Mirai is a malware that targeted IoT devices to create a very large botnet that was used for many distributed denial of service (DDoS) attacks in 2016. It did so by guessing the credentials of the devices. Many users did not change the credentials of the device from the factory defaults due to which it was very easy for the malware to infect hundreds of thousands of devices. An army of such devices was used to conduct DDoS attacks on companies like Krebs on Security, OVH, and Dyn. This alone highlights the disastrous risks of having vulnerabilities on IoT-based systems.

In addition to this, privacy concerns also arise when we have devices that can monitor our daily activities and transfer that data through a large network. Any device that is connected to the Internet is susceptible to many different forms of attack that involve stealing one's identity, accessing other people's data, and so on. Websites like Shodan (shodan.io) allow people all over the Internet to legitimately access other people's devices like IP cameras [22]. In this age, where privacy is of utmost importance, IoT devices have to maintain the highest level of privacy possible.

1.7 Conclusion

In this chapter, a brief overview of the entire IoT ecosystem was presented along with its history, applications, enabling technologies, and recent developments and challenges. IoT is a technology with vast applications in many fields that affect our day-to-day lives. Improvements brought in the fields of healthcare, industries, energy management, etc., by the sensing, monitoring, and communicating capabilities of IoT have greatly upgraded our lifestyle and it will keep on enhancing our lifestyle in the future as the IoT technology further evolves. However, IoT is still not a technology that is without faults. Many technical and ethical issues arising from the use of IoT have hindered it from its widespread adoption. However, IoT remains one of the most influential pieces of technology in the present world, and, thus, with ongoing research and developments, these limitations will hopefully be overcome in the near future. Therefore, IoT is a field with nearly limitless potential in the foreseeable future.

References

[1] K. Ashton, "That 'Internet of Things' thing," *RFID Journal*, 22-Jun-2009. [Online]. Available: https://www.rfidjournal.com/that-internet-of -things-thing. [Accessed: 07-Sep-2021].

[2] "The 'Only' Coke Machine on the Internet," Carnegie Mellon School of Computer Science. [Online]. Available: https://www.cs.cmu.edu/~{}c oke/history_long.txt. [Accessed: 07-Sep-2021].

[3] E. Dave, "The Internet of things: How the next evolution of the Internet is changing everything," 2011. [Online]. Available: https://www.cisco. com/c/dam/en_us/about/ac79/docs/innov/IoT_IBSG_0411FINAL.pdf [Accessed: 23-Oct-2021].

[4] C. Floerkemeier, E. Fleisch, M. Langheinrich, F. Mattern, and S.E. Sarma, in *Proceedings of the Internet of Things First International Conference, IoT 2008*, Zurich, Switzerland, March 26–28, 2008. Berlin, Heidelberg: Springer-Verlag Berlin Heidelberg, 2008.

[5] E. Irmak and M. Bozdal, "Internet of Things (IoT): The most up-to-date challenges, architectures, emerging trends and potential opportunities," *International Journal of Computer Applications*, vol. 179, pp. 20–27, May 2018.

[6] N. Sharma, M. Shamkuwar, and I. Singh, "The history, present and future with IoT," *Intelligent Systems Reference Library*, pp. 27–51, 2018.

[7] M. Weiser, "The computer for the 21st century," *ACM Sigmobile Mobile Computing and Communications Review*, vol. 3, no. 3, pp. 3–11, 1999.

[8] R. De Michele and M. Furini, "IoT healthcare," in *Proceedings of the 5th EAI International Conference on Smart Objects and Technologies for Social Good*, 2019.

[9] K. Saleem, I.S. Bajwa, N. Sarwar, W. Anwar, and A. Ashraf, "IoT healthcare: Design of smart and cost-effective sleep quality monitoring system," *Journal of Sensors*, vol. 2020, pp. 1–17, 2020.

[10] S. Allen, "2021 global health care sector outlook," *Deloitte*, 25-Jun-2021. [Online]. Available: https://www2.deloitte.com/global/en/pa ges/life-sciences-and-healthcare/articles/global-health-care-sector-outl ook.html. [Accessed: 07-Sep-2021].

[11] M. Ersue, D. Romascanu, J. Schönwälder, and A. Sehgal, "Management of networks with constrained devices: Use cases," no. 7548. RFC Editor, May 2015.

[12] P.M. Kumar and U. Devi Gandhi, "A three-tier Internet of Things architecture with machine learning algorithm for early detection of heart diseases," *Computers & Electrical Engineering*, vol. 65, pp. 222–235, 2018.

[13] R.N. Kirtana and Y.V. Lokeswari, "An IoT based remote HRV monitoring system for hypertensive patients," in *Proceedings of 2017 International Conference on Computer, Communication and Signal Processing (ICCCSP)*, 2017, pp. 1–6, doi: 10.1109/ICCCSP.2017.7944086.

[14] W. Li, Y. Chai, F. Khan, S.R. Jan, S. Verma, V.G. Menon, Kavita, and X. Li, "A comprehensive survey on machine learning-based big data analytics for IoT-enabled smart healthcare system," *Mobile Networks and Applications*, vol. 26, no. 1, pp. 234–252, 2021.

[15] T. de Vass, H. Shee, and S. Miah, "IoT in supply chain management: Opportunities and challenges for businesses in early Industry 4.0 context," *Operations and Supply Chain Management: An International Journal*, pp. 148–161, 2021.

[16] T.M. Brown-Brandl, "1 using RFID in animal management and more," *Journal of Animal Science*, vol. 97, no. Supplement_2, pp. 1–2, 2019.

[17] S. Misra, A. Mukherjee, and A. Roy, *Introduction to IoT*. Cambridge: Cambridge University Press, 2021.

[18] H. Park, H. Kim, H. Joo, and J.S. Song, "Recent advancements in the Internet-of-Things related standards: A onem2m perspective," *ICT Express*, vol. 2, no. 3, pp. 126–129, 2016.

[19] M. Noura, M. Atiquzzaman, and M. Gaedke, "Interoperability in internet of things: Taxonomies and open challenges," *Mobile Networks and Applications,* vol. 24, no. 3, pp. 796–809, 2018.

[20] M. Aly, F. Khomh, Y.-G. Gueheneuc, H. Washizaki, and S. Yacout, "Is fragmentation a threat to the success of the internet of things?," *IEEE Internet of Things Journal*, vol. 6, no. 1, pp. 472–487, 2019.

[21] M. Antonakakis *et al.*, "Understanding the Mirai Botnet," in *Proceedings of the 26th USENIX Conference on Security Symposium*, 2017, pp. 1093–1110.

[22] H. Lin and N. Bergmann, "IoT privacy and security challenges for smart home environments," *Information*, vol. 7, no. 3, pp. 44, 2016.

[23] *Alexa Smart Home-Learn About Home Automation*. [Online]. Available: https://www.amazon.com/alexa-smart-home [Accessed: 23-Oct-2021].

[24] F. Wang, L. Hu, J. Hu, J. Zhou, and K. Zhao, "Recent advances in the internet of things: Multiple perspectives," *IETE Technical Review*, 2016, doi: 10.1080/02564602.2016.1155419.

[25] Available: https://pixabay.com/illustrations/iot-internet-of-things-network-3337536/ [Accessed: 23-Oct-2021].

[26] A. Sarac, N. Absi, and S. Dauzère-Pérès, "A literature review on the impact of RFID technologies on supply chain management," *International Journal of Production Economics*, vol. 128, no. 1, pp. 77–95, 2010.

[27] M. Attaran, "Digital technology enablers and their implications for supply chain management," *Supply Chain Forum: An International Journal*, vol. 21, no. 3, pp. 158–172, 2020.

[28] M.T. Okano, "IoT and industry 4.0: The industrial new revolution," in *Proceedings of International Conference on Management and Information Systems*, vol. 25, Sep. 2017, pp. 26.

[29] L.D. Xu, W. He, and S. Li, "Internet of things in industries: A survey," *IEEE Transactions on Industrial Informatics*, vol. 10, no. 4, pp. 2233–2243, Nov. 2014, doi: 10.1109/TII.2014.2300753.

[30] V. Marinakis and H. Doukas, "An advanced IoT-based system for intelligent energy management in buildings," *Sensors*, vol. 18, no. 2, pp. 610, Feb. 2018. [Online]. Available: http://dx.doi.org/10.3390/s18020610

[31] J. Yun, S.-S. Lee, I.-Y. Ahn, M.-H. Song, and M.-W. Ryu, "Monitoring and control of energy consumption using smart sockets and smartphones," *Communications in Computer and Information Science*, pp. 284–290, 2012.

[32] K. Chooruang and K. Meekul, "Design of an IoT energy monitoring system," in *Proceedings of 2018 16th International Conference on ICT and Knowledge Engineering (ICT&KE)*, 2018, pp. 1–4, doi: 10.1109/IC TKE.2018.8612412.

[33] R.R. Hernandez, S.B. Easter, M.L. Murphy-Mariscal, F.T. Maestre, M. Tavassoli, E.B. Allen, C.W. Barrows, J. Belnap, R. Ochoa-Hueso, S. Ravi, and M.F. Allen, "Environmental impacts of utility-scale solar energy," *Renewable and Sustainable Energy Reviews*, vol. 29, pp. 766–779, 2014.

[34] A.S. Spanias, "Solar energy management as an internet of things (IoT) application," in *Proceedings of 2017 8th International Conference on Information, Intelligence, Systems & Applications (IISA)*, 2017.

[35] S. Rao *et al.*, "An 18 kW solar array research facility for fault detection experiments," in *Proceedings of 2016 18th Mediterranean Electrotechnical Conference (MELECON)*, 2016, pp. 1-5, doi: 10.1109/MELCON .2016.7495369.

[36] M. Sharma, M.K. Singla, P. Nijhawan, S. Ganguli, and S.S. Rajest, "An application of IoT to develop concept of smart remote

monitoring system," *Business Intelligence for Enterprise Internet of Things*, pp. 233–239, 2020.

[37] D.D. Prasanna Rani, D. Suresh, P. Rao Kapula, C.H. Mohammad Akram, N. Hemalatha, and P. Kumar Soni, "IoT based smart solar energy monitoring systems," *Materials Today: Proceedings*, 2021.

[38] M. Ali and M.K. Paracha, "An IoT based approach for monitoring solar power consumption with Adafruit cloud," *International Journal of Engineering Applied Sciences and Technology*, vol. 04, no. 09, pp. 335–341, 2020.

[39] P. Siano, "Demand response and smart grids—A survey," *Renewable and Sustainable Energy Reviews*, vol. 30, pp. 461–478, 2014.

[40] V.C. Gungor, B. Lu, and G.P. Hancke, "Opportunities and challenges of wireless sensor networks in smart grid," *IEEE Transactions on Industrial Electronics*, vol. 57, no. 10, pp. 3557-3564, Oct. 2010, doi: 10.1109/TIE.2009.2039455.

[41] M. Anjanappa, K. Datta, T. Song, R. Angara, and S. Li, "Introduction to sensors and actuation," in *The Mechatronics Handbook*, 2nd ed., R. H. Bishop, Ed. Hoboken: CRC Press, 2008.

[42] H. Janocha, Ed., *Actuators*, 1st ed. Berlin: Springer, 2004.

[43] P.K.D. Pramanik and P. Choudhury, "IoT data processing: The different archetypes and their security and privacy assessment," in *Internet of Things Security: Fundamentals, Techniques and Applications*, S. K. Shandilya, S. A. Chun, S. Shandilya, and E. Weippl, Eds. Gistrup: River Publishers, 2018, pp. 37–54.

[44] A. Rayes and S. Salam, "The Internet in IoT—OSI, TCP/IP, IPv4, IPv6 and Internet routing," in *Internet of Things from Hype to Reality*. Berlin: Springer, 2017, pp. 35–56.

[45] H. Zimmermann, "OSI reference model—The ISO model of architecture for open systems interconnection," *IEEE Transactions on Communications*, vol. 28, no. 4, pp. 425-432, Apr. 1980, doi: 10.1109/TCOM.1 980.1094702.

[46] S. Al-Sarawi, M. Anbar, K. Alieyan, and M. Alzubaidi, "Internet of Things (IoT) communication protocols: Review," in *Proceedings of 2017 8th International Conference on Information Technology (ICIT)*, 2017, pp. 685–690, doi: 10.1109/ICITECH.2017.8079928.

[47] A. Balan, T. Balan, M. Cirstea, and F. Sandu, "A PUF-based cryptographic security solution for IoT systems on chip," *EURASIP Journal on Wireless Communications and Networking*, vol. 2020, no. 1, 2020.

[48] G. Hu, W.P. Tay, and Y. Wen, "Cloud robotics: architecture, challenges and applications," *IEEE Network*, vol. 26, no. 3, pp. 21-28, May-Jun. 2012, doi: 10.1109/MNET.2012.6201212.

[49] Available: https://www.freesvg.org [Accessed: 23-Oct-2021].

[50] D.C Motor image by Dcaldero8983, CC BY-SA 3.0. Available: https://creativecommons.org/licenses/by-sa/3.0, via Wikimedia Commons.

[51] Images by SparkFun, CC BY 2.0. Available: https://creativecommons.org/licenses/by/2.0

[52] PIR image by Nowforever CC BY-SA 4.0. Available: https://creativecommons.org/licenses/by-sa/4.0 via Wikimedia Commons.

[53] O. Alfandi, S. Khanji, L. Ahmad, and A. Khattak, "A survey on boosting IoT security and privacy through blockchain," *Cluster Computing*, vol. 24, no. 1, pp. 37–55, 2020.

[54] S.N. Mohanty, K.C. Ramya, S.S. Rani, D. Gupta, K. Shankar, S K. Lakshmanaprabu, and A. Khanna, "An efficient lightweight integrated blockchain (ELIB) model for IoT security and privacy," *Future Generation Computer Systems*, vol. 102, pp. 1027–1037, 2020.

[55] B.K. Mohanta, D. Jena, U. Satapathy, and S. Patnaik, "Survey on IoT security: Challenges and solution using machine learning, artificial intelligence and blockchain technology," *Internet of Things*, vol. 11, 2020, Art. no. 100227.

[56] R. Ahmad and I. Alsmadi, "Machine learning approaches to IoT security: A systematic literature review," *Internet of Things*, vol. 14, 2021, Art. no. 100365.

[57] M.A. Amanullah, R.A. Habeeb, F.H. Nasaruddin, A. Gani, E. Ahmed, A.S. Nainar, N.M. Akim, and M. Imran, "Deep learning and big data technologies for IoT security," *Computer Communications*, vol. 151, pp. 495–517, 2020.

[58] W. Ahmed *et al.*, "Machine learning based energy management model for smart grid and renewable energy districts," *IEEE Access*, vol. 8, pp. 185059–185078, 2020, doi: 10.1109/ACCESS.2020.3029943.

[59] P. Sivaraman and C. Sharmeela, IoT-Based Battery Management System for Hybrid Electric Vehicle, in P. Sanjeevikumar (eds), Artificial Intelligent Techniques for Electric and Hybrid Electric Vehicles, John Wiley-Scrivener Publishing, 2020.

2

IoT and its Requirements for Renewable Energy Resources

D. Gunapriya[1], R. Sivakumar[2], K. Sabareeshwaran[3], and C. Sharmeela[4]

[1]Sri Krishna College of Engineering and Technology, India
[2]Akshaya College of Engineering and Technology, India
[3]Karpagam Institute of Technology, India
[4]Anna University, India
E-mail: gunapriyadevarajan@gmail.com; Sivakkumar14@gmail.com;
sabareeshwarank@gmail.com

Abstract

The worldwide energy production is majorly based on fossil fuels, and it covers 84% of world key energy consumption as on 2020 statistics. To support global energy demand, both government and private sectors have started to focus the problems on fossil fuel depletion and the impact of Greenhouse gas emissions in climatic change. To reduce the emission of Greenhouse gas and to obtain energy transition of sustainable range, it is necessary to integrate renewable energy in the power grid as per recent studies. Though the renewable energy sector kept its phenomenal growth in the current scenario and the prediction of its supply to fulfill two-thirds of global energy demand in 2050, there are challenges in power generation, transmission, and distribution where the economic feasibility plays a role. Modern technology, such as the Internet of Things (IoT), plays an important role in overcoming these problems and optimizing renewable energy supplies. Understanding the use of IoT technology fosters trust in this sector to give a better solution for renewable energy management such as generation, transmission, and distribution, as well as effective energy sharing to the grid. This chapter discusses IoT technologies and their integration in the renewable energy sector to increase their efficiency. Furthermore, information transmission and storage for data analytics to provide optimum energy supply, and issues in IoT implementation with solutions to security and privacy maintenance are highlighted.

2.1 Introduction

2.1.1 IoT and its Necessity

The modern equipment offers effective communication among themselves besides carrying out their functions, which supports effective system operation. In the fourth industrial revolution, robotics systems, wireless communication, the Internet of Things (IoT), cloud computing, and so on dominated the functions [1]. Now the issue arises as to "Why should the equipment in the system, or system to system, interact with each other?" The solution is that by using effective communication, superfluous functions in the system may be turned off to conserve energy and optimize system operations. Furthermore, the gathering of operational information (equipment data and environmental data) in real time via communication is utilized for data analytics to predict/measure equipment performance to avoid unexpected system failure [2]. In addition, data analysis is utilized to make decisions on system enhancements and adjustments depending on future demand. For more than a decade, machines interacted using machine-to-machine (M2M) communication, a one-to-one communication method used to link one machine to another [3]. However, when sensors and actuators are utilized with devices for sending and receiving data to monitor the devices, transmission control protocol/Internet protocol (TCP/IP) is used for large information transmission between the communicating devices. Kevin Ashton invented the term "Internet of Things" (IoT) in 1999, and his concept of IoT technology was based on radio frequency identification (RFID) device communication, which differs greatly from today's IP-based connectivity. After 2011, the IoT boom was happening with the network layer of IPv4 and IPv6, which became essential to IoT with the aid of developing technologies. In this rapid development, there are already 6.4 billion linked devices and sensors in IoT, with 5.5 million devices connecting to IoT on a daily basis [4].

2.1.2 Challenges in RES

The current fluctuating utility power demand in the distribution system worldwide is managed by incorporating the variable renewable energy resources (VRES) into the distribution system with the help of decarbonizing effect.

Table 2.1 RES challenges as per the root cause analysis.

Factors	Cause
Stability	Stability violations and re-dispatch allocation
Steadiness	The mismatch between source and load
Eminence	Safety hazards and aging of equipment

VRES, such as solar power grid systems and wind power grid systems, plays a significant role in accomplishing the power sector decarbonization. However, they are not the same as traditional power production systems. By integrating the VRES in the distribution system, there will be possibilities for the formation of difficulties and challenges in the operation of the distribution network (DN) [5]. Furthermore, in the context of the growing role of VRES in meeting current power demand, it is, therefore, imperative to address the new problems that arise in power systems. If these problems and issues are not addressed, they may endanger the reliability of the power supply system as well as the decarbonization targets [6].

Table 2.1 lists RES challenges as per the root cause analysis and offers an overview of the signs of growing VRE penetration discovered by a literature study. The symptoms may be classified that correspond to the power system's essential performance criteria. The categories are briefly described in the sections that follow. The end user's primary performance requirement is adequate power quality. The power quality factor includes the conditions for uninterrupted power sources, consistent voltage and current parameters, and safe conditions in the event of disruptions. The uniqueness and asynchronous facts of VRES are two basic features that are primarily responsible for power quality issues. The current factor is concerned with the effective DN of electricity. In comparison to the other categories, the root causes of problems in the flow category are numerous. The greatest proportion of flow problems is caused by VRE unpredictability, modularity, and location restrictions. The stability factor is concerned with power rate modulations and voltage fluctuation, together with restart and recovery of the system after the power distribution interruptions. The problems that deal with the balancing of the power supply with short and long terms in the distributions systems as well as managing the demand are the concerns of the eminence factor in the VRES power generation system. This includes the management of the power systems' upgradation competency and maintaining low power generation levels among the systems. But VRES power generation fluctuations and unpredictability create balancing problems. To summarize, the analysis of causes in VRES stipulates a reliable foundation for the classification of problems in power systems with its growth and adaptations.

2.1.3 Integration of IoT in RES and Benefits

Due to the depletion of fossil fuels and the growth in environmental contamination, contemporary technological development and research have mostly focused on the adoption of green energy production sources and the best use of existing energy resources [7]. Smart grid and microgrid technology innovation is highly useful in achieving these aims through power suppliers. The purpose of smart grids is to effectively deal with energy distribution management which involves the generation, transmission, and distribution operations. To effectively manufacture, use, and distribute energy, all participants, such as distributors, customers, and producers, are authorized to engage in two-way communication. As a result, energy efficiency may be achieved by effectively using the available energy sources. Since smart and microgrid technologies are becoming a preferred method to power generation due to the focus on green energy production globally, there is also an increase in demand for cleaner energy manufacturers and suppliers.

The recent research works depict that the recent developments of sensors, data storage systems, and data analytical tools are majorly used to achieve the balancing of power supply against the power requirement effectively to the consumers who are connected to the smart grid system. The advancements of smart devices and their connectivity through the Internet bring the meaning of the Industrial Internet of Things (IIoT) and the purpose of use in power grid system. Smart devices aid in increasing operational efficiency, optimizing company operations, and safeguarding the system. Power generation planning is essential to optimize the usage of power and its costs. The energy planning must include the numerous energy resources available and then optimize depending on feedback from many associated elements. The energy consumption needs are conveyed to the framework via the IoT, which aids in communication with various IoT-based smart devices and provides a response to smart grids to achieve the optimal decision.

2.2 Industrial IoT

In today's world, the rapid development of IoT is radically transforming people's lives. Currently, IoT manufacturers are concentrating their efforts on offering IoT solutions to specific industries rather than broad applications. There is also a significant opportunity in the rising demand for IoT devices in the industrial sector to deliver industry-specific solutions. The IIoT is distinguished by the interconnection of machines, people, and computers to enable smart industrial processes that use advanced data analytics to alter business outcomes. The IIoT is an extension of the IoT that has evolved as a broad notion of applying the IoT to the industrial sector [8].

However, because of the integration requirement, it confronts specific problems that distinguish it from other IoT systems and services. Although the fundamental concepts of IIoT and IoT are similar, i.e., smart device connections that enable remote operation of the equipment, receive and store data, analyze the data, and supervise and control the variables that distinguish the IIoT as the subcategory of IoT which develops the strict requirements for uninterrupted services and security with efficient operational technology used in the industrial fields. The industrial sector's particular qualities - technology and requirements – result in specialized solutions and services to support the industrial division's concentration on a customized IoT concept. As a result, the industrial sector has taken a keen interest in the creation of specialized ideas ranging from policy to technology services.

2.2.1 Architecture of IoT

The first stage in building an IoT system for a given application is to select acceptable and dependable IoT devices, communication protocols, data storage devices, and data analytic methodologies for application performance prediction [9]. The advancement and distribution of IIoT systems and services require the creation of designs that allow for competent and active operations, including communication, predicted endway services, and the major investors tangled for devices, cyber–physical systems, communication networks (CNs), service providers, and professional promoters. The International Telecommunication Union (ITU) addressed the issues in the development of standards and architecture for IoT by publishing the ITU-T Y.2060 recommendation in 2012, which presents a reference architecture for IoT in general as well as applications that come under the IIoT umbrella, such as smart grid system and intelligent transportation systems, e-wellbeing, and so on. The ITU initiative has broadened the communications vision to encompass communication of "everything" as well as communicate ideas of "any time" and "any place." It is vital to highlight that it covers all anticipated applications, including industrial ones like smart grids and smart transportation systems, among others. ITU considers physical and virtual items that are recognizable and capable of connecting to CNs to be "things." Importantly, because communication is such a key component of the IoT concept, physical items that fall under the "devices" category must be connected to networks for any analog data to be converted to digital and communicated via Internet connections. Devices can be simple data transmitting devices that transfer

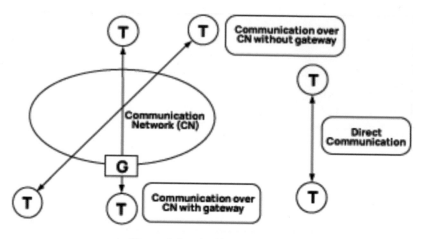

Figure 2.1　Architecture of IoT.

and store data, data acquisition devices that interact with the equipment via encoders and decoders, sensors and actuators or machines, end service devices, and consumer electronic devices that have embedded processing and communication resources.

The connectivity of devices is the fundamental idea of IoT architecture based on the ITU reference model. Figure 2.1 shows the architecture of IoT wherein the model analyzes three communication techniques based on the usage of gateways (G), local network (LN), and the CN. The smart devices can connect openly, over LNs, and/or over the CN, without the need for gateways, or they can communicate via gateways. The IIoT has the key features of interconnectivity, scalability, diversity, service facility, the dynamic information of the devices, and connectivity. Development becomes a substantial factor that needs to be included at all levels of IIoT as the number of connected devices increases dramatically. The development issue is present not only to communication points and the number of smart devices but also to the size of collected and transmitted data including its administration in terms of huge storage and analyzing. The dynamic performances of the devices, which turn on and off automatically or join and detach from CNs, will complicate and demand more from the landscape.

2.2.2 IoT Components

Figure 2.2 illustrates the IIoT-ITU reference model, which was presented to satisfy the aforementioned needs. It is a normal layer construction model

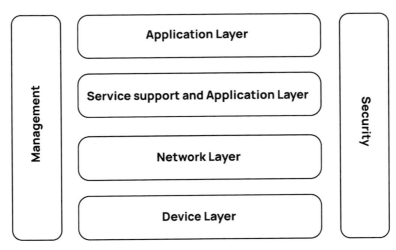

Figure 2.2 IIoT-ITU reference model.

which consists of four tiered layers, particularly device layer, network layer, software and service assistance layer, and lastly application layer, and two vertically integrated layers which are crosscutting the four-tiered layers, defining managing as well as security capabilities and attributes to all hierarchical levels. The first layer that presents the device information, probably the bottom in the hierarchy, consists of the performance of equipment and CNs. Thinking about the primary interest of ITU in marketing communications, the level describes communication-centered performance for the devices: Importantly, the unit level includes protocol transformation because products might put into action different protocols and, therefore, requires process transformation for interoperability.

The network layer provides a summary of related protocols and the conversion of the data from the devices to network-level protocols. This layer contains information about how the network works, including how the data in the model is transported. For networking, they may include control options for accounting, authorization, authentication, mobility, and network connectivity, while for transportation, they foresee user traffic transportation along with the transport of management and management info for (I)IoT service as well as applications. The service assistance and application assistance layers include both service/application-specific and generic functionality (capabilities) which allows (I)IoT applications as well as services. Looking at the distributed character of (I)IoT services as well as programs, there exists generic efficiency, like information processing and storage and

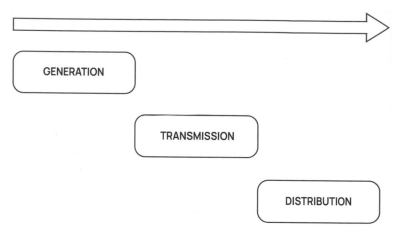

Figure 2.3   Flow of electrical energy in conventional electric system.

particular functionality per service and application since emerging solutions have requirements that are different; for instance, smart grid operation puts various privacy needs than a smart toll management process for transportation services. Lastly, the application level, probably the highest hierarchical level, consists of the (I)IoT applications as well as services. The managing vertical and crosscutting level involves both generic and program domain-specific functionalities, account management, security, fault, performance, and resource.

2.3 RES and IoT

For a long time, the current electrical grid design has provided excellent service to anyone who has access to energy. Even though the unit is evolving, it is critical to comprehend the devices that have brought us to this point. The facility of electrical energy typically alters on three important segments: distribution, transmission, and generation. Figure 2.3 shows the ?ow of electrical energy in the conventional electric system.

Approximately 5% of electrical energy is lost as line losses in long-distance transmission lines. The existing power grid was meant to satisfy the twentieth-century demand.

2.3.1 IoT Controls for RES

The IoT is used to transmit the energy usage requirements to the framework, which allows it to interface with various IoT-based devices and send feedback to microgrids to carry out the optimal option. Four goals could be achieved using the projected look as follows.

- Initially, it creates a self-contained IoT for transmitting and receiving data from end-users.
- As a result of the foregoing, it regulates the power usage of products based on the goals and constraints established by customers.
- Third, it provides buyers with feedback on their power consumption patterns to help them save money and energy.
- Finally, it has inexhaustible power sources in the network.

It provides applications like remote monitoring, prognostic maintenance, and cybersecurity and enhances the professional process, laborer well-being, and then advanced distributed management.

To enhance the consumption of distributed power resources (DERs), the grid must become more intelligent to improve traditional grid infrastructure while also promising IoT coordination. Transmission operators can account for the distribution and generation elements of the grid's resources and assets. Operators can regulate crucial aspects of the grid while also including distribution. A decentralized strategy brings the generation near to the needed load, reduces transmission losses as well as vulnerabilities, and expands the overall dependability, versatility, efficiency, and sturdiness of the grid. Communication is closer and bidirectional to near real-time, empowering customers to be apt to oversee expenses and loads. Additionally, power fees may be progressively unique. Besides, smart building products empower intelligent equipment through the web. The smart grid bolsters the use of even more nuanced and effective demand management plans as well as the delivery of progressively educated methods by the customers. Intelligent digital meters, also referred to as "smart meters," are basic to the smart grid. To enhance the utilization of unlimited energy resources, these grids improve customer involvement infrastructure as well as assurance to sign up with IoT at the community level. On a town level, a reasonable microgrid allows for a bidirectional stream of communication and electricity between electricity suppliers and customers. IIoT plays a critical role in addressing the issue of non-accessibility of specific renewable energy solutions by monitoring power use, energy generation, and integration, particularly for intelligent microgrids. Smart microgrids improve nearby dependability by utilizing an explicit dependability enhancement program that includes surplus distribution, clever switches, energy generation, energy storage, automation, and other smart technologies. With the advances and switches in the electricity sector, utilities, as well as the power sector, constantly change for the long term. Nevertheless, the grid is going to keep on being essential to the electricity industry.

2.3.2 Challenges in IoT Implementation

The basic functions of smart grid (SG) advancement consist of the development of power grid systems, standards and communication, analytical intelligence, economic and environmental factors, and testbed. Several of the difficulties for sustainable growth include electricity issues, manpower capabilities, political policies, and technology [10]. The computational intelligence attribute consists of innovative analytical equipment that will enhance the majority power network by using heuristics, evolution programming, choice support equipment, and adaptive optimization methods, which are promising resources for the layout and computation demand of SG. Renewable energy development makes use of variability, and its consumption is also commercially and economically feasible to alleviate demand anxiety while also improving reliability, minimizing losses, and reducing the carbon footprint caused by winter and gas-based energy solutions. The improvement of automation, interaction, and requirements is needed to make sure that rapid decision-making promotes effectiveness and dependable operation. These options are determined by the energy and the customers. Human development, homegrown know-how, the work of scholars and government, interconnection challenges, engineering challenges, technologies in plant layout, maintenance, and operation, and improvement of SG with notable architecture for supporting energy efficiency as well as need for supply control are pathways for the renewable development of SG.

Generally, the primary technical challenge in the strength process functioning in a deregulated or competitive atmosphere is increasing the power transfer power of current transmission methods to stay away from the congestion in the method. Many techniques have been recommended to manage the problem. Several of them include maximum energy flow, grounded model scheduling, use of complex systems like FACTS, and sending out decades, and they will help you mitigate congestive network situations on the constrained transmission track. The issues for consideration in computational complications for the improvement of the SG feature are the following:

- penetration of unlimited energy resources (RERs); participant bidding methods prevent businesses from providing answers with environmental objectives;

- the high reliance on energy method versions on intelligent operation and influence, power method planning, and management prevents businesses from providing solutions with complex goals;
- lack of comprehensive knowledge by engineers and operators of computational equipment that is actually user-friendly and immediately interpretable;
- the control and operation are complex power systems because of the complexity of computational resources used for uncertainties and modeling;
- forecasting tonne demand, ancillary services, and prices avoids businesses from providing answers with economic objectives;
- risk minimization in the power sector that is electric with buy-in computational resources seeks to establish a tradeoff between maximizing the anticipated return shipping.

The main complex issues of SG are interoperability, networking communications, demand response, power storage, and distribution grid control.

2.4 Challenges of IoT in EMS Post-implementation

The challenges of IoT are classified into various categories as stated in the following:

- privacy issues;
- security issues;
- data storage issues.

2.4.1 Privacy Issues

The usage of IoT is widespread as it offers a unique operation for various applications such as wellness and transportation systems. IoT also provides a better possibility for the interconnection of various sensors. This system has some privacy issues, such as the following.

Authentication:

Authorization to access the database of IoT is the main privacy issue. This challenge arises as many rely on weak passwords for authenticating the IoT.

Encryption:

Encryption is the process of hiding secure data. It has been noted that many IoT-empowered gadgets do not encrypt the data while transferring over the Internet.

Insecure Interface:

IoT devices mostly have a screen for controlling or assigning the schedule. This interface needs proper authentication to alter the operation.

Credentials:

Many people incorporate and use the default credentials due to a lack of knowledge in changing the username and password. This is vulnerable to hackers accessing and re-altering the operation.

Coding Practice:

The IoT devices are designed to be user-friendly so that the machine coding can be changed as needed. But this has also been a challenge in privacy concerns since the coding is done or altered without the proper coding ethics.

User Privacy:

When transitioning to advancement, the privacy of the customer/user becomes much more important. For example, a grocery shop collects the user's information, particularly their location and the frequency of products purchased. Unless the data is securely handled or encrypted, the health of the user is exposed, which is a privacy thread concerning IoT.

Reducing Privacy Issues and Security Challenges:

IoT products are designed in such a way to be secure if the proper protocols are followed.

Network Validation:

Proper scheduled validation needs to be done, such as network traffic and any modifications in encryption. If the encryption is properly done, the security issues can be reduced to some extent.

Edge Channel Validation:

With the aid of penetration assessment, the edge channel defense can be validated for both software and hardware modules. This penetration will reduce the recent threads on privacy issues.

Coding Ethics:

The codes of various products are kept secure to be utilized for marketing purposes. Early safe code feedback is recommended to reduce the upcoming challenges. On the other hand, the financial cost of rectifying the security issues will be reduced if proper handling of the code is done at proper intervals.

2.4.2 Security Concerns

IoT-based systems are predicted to face various security threats, such as the following.

Spoofing:

It is like a misusage of a person's or company's identity. Spoofing aims to represent the identity in a different manner and to spoil the reputation.

Tampering:

Data tampering is mostly seen in the marketing category. This challenge is put forth by many leading organizations, such as obtaining and analyzing the usage of a particular product by consumers and increasing the price of that product. The tampering is not limited to organizations; it is done in residential too by obtaining the power usage from smart meters and imposing additional prices for the same usage.

Compromising:

It has been noted that only some products are tamper-resistant, whereas many fall under the non-tamper criteria. As an example, the meters are tamper-resistant, whereas the connected sensors or actuators are not tamper-resistant. Compromising the resistance will also lead to the security thread. Most of the IoT devices are IP enabled; so it is easier to incorporate into a common Internet. This methodology gives a key for attackers to access the data, leading to a security vulnerability. This cyber-attack not only damages the software part but also damages the physical parts of the IoT.

Scalable Issue:

The protection of the IoT is not scaled down to certain regions; it has been designed to cover a wide range of products. This is the key for attackers to easily obtain the credentials of the user.

IoT Communication:

Nowadays, the use of driverless cars is common. The technology behind this is IoT. If proper security is not provided, it may lead to unpredictable damage to both humans and the product.

Deployment:

The SG is intended to cover a wide area, as a single SG covers the entire nation, which leads to the work being unsupervised. If adequate SGs are provided at distant locations without any physical boundary, monitoring security is easily accessible and tampering of data can also be mitigated to some extent.

Outdated Systems:

The integration or connection of IoT security monitoring is done in the old available system and also through some private bodies. Usage of IoT-based SG in the existing system is a noteworthy issue since there is no possibility to connect the existing system with the new upgraded modules.

Resources:

Most of the SG devices are restricted in accessing the resources, which are predominantly widely connected. For enhancement of the security concerns, particular attention is needed to accommodate the solutions.

Integration:

There is still usage of legacy systems that do not support the TCP/IP protocol and cannot connect with IP-based systems without the aid of third-party gateways, leading to a lack of end-to-end encryption. And also, there is a notable issue of security-enhanced device interconnection with unsecured devices, which makes both the devices vulnerable to security issues.

Trust Management:

Devices that are from different entities do not acquire trust till a minimum level of security enhancement is provided. Nonetheless, the question of establishing trust between various entity device connections remains unanswered, particularly in long-distance connections.

Bandwidth:

The SG devices are to be made to respond faster and more accurately to time-to-time variations. SCADA, for example, is a computer-controlled system used in transmission and distribution systems that must be capable of responding to time-to-time variations in electrical parameters such as voltage, current, power, and so on. The devices employed in weather prediction should be very fast in updating the meteorological changes by obtaining data from various sensors.

Enhancement of IoT Security:

The key security that needs to be incorporated for the IoT used in the smart grid is as listed in the following.

Validation:

The competence to check and verify the actual devices connected and detect if there is any new unknown device connected or accessing the data transmission and distribution. The energy supplier needs to validate the user's smart meter for consolidating the bill of use.

Genuineness:

The supplier needs to provide a security enhancement to the database so that the accessed data or bill does not tamper illegally. Periodic software enhancement is needed for smart meters, as the whole process relies on the connected meter.

Inscrutability:

The secured data of the user should be made accessible only by the allocated SG operator and the energy provider for that particular user. And, additionally, the user is restricted to knowing only the particular data accessed and not the whole accessed data of all consumers.

Privacy Protection:

Even though the SG operator and energy provider have the data of the user, irrespective of the units consumed or energy price, the data should not be used for any other purpose than the calculation procedure of the bill. Simply, the data collected for billing should not be used for another purpose without proper approval from the user.

Authentication:

Ensure that the allocated person from the SG is accessing the data on both software and hardware. Periodic validation is required to ensure that the allocation of rights for data access is needed. If any mismatch is noted in the usage of power, the authorized person needs to make a field visit and investigate whether there are any configuration issues in the smart meter or any physical alterations have been made.

2.4.3 Data Storage Issues

2.4.3.1 Challenges in data management

Every IoT device is capable of collecting a large amount of data. The issue occurs in the storage of all the data at a single point. As per the research, it has been clear that, at present, there is no adequate storage available to store the entire sensor and actuator collected data. Very few entities have invested in developing an adequate storage space for housing all their device data. Many have compromised on storing all the data by storing only the data based on the need and its value. But this causes an issue when we need data, which, at that time, may be considered unneeded. Increasing the number of data centers will only be a part of the solution to deal with the daily increase in IoT device usage. The proper segregation of data from the device is needed so that it is only easily accessible with available bandwidth. For instance, if we need to calculate the usage of power by a user, the accessing of smart meter details should be simplified, and, most of the time, they need to be quickly accessible.

2.4.3.2 Challenges in fetching data

Even though all the data is stored in data management, for easy access, we rely on data fetching tools. The data stored is not only the initial cases but also needs periodic updating. For example, the billing details of the user need to be updated at every instance of calculating the energy usage. The streaming of weather data is based on the data obtained from the sensor and stored in a particular location/allocation of the dataset, with the data being fetched from a particular set from time to time, for instance. Data fetching tools play a significant role in updating the proper data to the allocated location and fetching the required data from the correct dataset. This fetching also needs to be done as per the individual user's needs. The streaming of digital content is made simple with these data mining tools. They have been enhanced in terms of security to detect tampering and illegal data retrieval. The digital media streams are completely reliant on these tools, as significant progress has been made in the data retrieval process.

2.4.3.3 Challenges in allocation

When comparing the development in consumer-related products, the development of IoT has a significant mark, such as innovation in sensors and wireless technologies. The day-to-day enhancement also needs to be under the security protocol. New IoT innovations will be unmotivated if they deviate from security concerns and standards. A small mistake in data handling can lead to unpredictable damage. For example, when handling the SCADA system, if any wrong data has been put forward based on that data, modifications are made to the physical devices, which leads to damage to the user-connected devices. Another instance, such as health monitoring, where wrong data about health is interpreted and the user consumes medicine concerning the IoT data, leads to unpredictable damage to humans. If even one of these incidents is reported and made public, users' trust in IoT devices will suffer. The data management needs to fetch and provide the user-required data irrespective of the available bandwidth. The issue occurs when there is saturation on the Internet since most of the IoT devices are operated on IP. The usage of IoT is fulfilled only when every user has trust and enhancement in the security of the system.

2.5 Solution to IoT Challenges

Through the implementation of IoT-based technology in the smart grid, it has become more reliable and efficient in terms of operation. Even though IoT provides good support in the operation of the smart grid system, it has certain security flaws, as discussed in previous topics. Those security issues can be overcome with the aid of various methodologies such as blockchain, artificial intelligence based operations, and distributed frameworks. In the upcoming sections, certain opportunities available in the IoT to mitigate security issues are discussed.

2.5.1 Blockchain Methodology

The blockchain methodology is an upcoming methodology concerned with the distribution of information and the management of storage. The major advantage of blockchain is that there is no centralized administration. In the context of IoT with blockchain, it has the capability of interconnection as well as information exchange in a peer-to-peer (P2P) community, as shown in Figures 2.4(a) and (b), which depict information exchange in a client-server network and a P2P network, respectively. Blockchain technology operates based on a centralized algorithm; so there is less possibility of data access by any third person. If any information on the server is accessed or modified, that info is shared with all the community users, which enhances the security of the smart grid data.

Blockchain technology has a periodic update in terms of enhancing security, which makes it a failure for cybercriminals to access centralized data. And also, blockchain integration with IoT is done in a manner such that the data can be backtracked and also notified if there is any conflict in the provided information. The data and information in the IoT are encrypted through blockchain engineering. In the P2P communication process, the interactions are built with the aid of a key exchange mechanism [11].

Through the cryptographic key exchange mechanism, the blockchain-protected data is assured to be safe and no intruder can interface with and

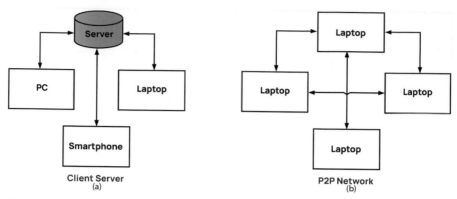

Figure 2.4 (a) Exchange of information in client-server network. (b) Exchange of information in P2P network.

access the data. The IoT-based smart grid data can be easily backtracked consequently. In the next part, the major function of blockchain, which benefits the IoT-implemented smart grid, is discussed.

- Nature of Distribution:
  The distribution process is faster, reliable, and efficient due to the elimination of various centralized architectural issues.
- Security Protection:
  Since all the data transmission is encrypted cryptographically, the reliability in the privacy protection is enhanced [12].
- Eminence:
  The possibility of data backtracking helps in analyzing all the changes or modifications made if any in the system. Easy traceability of blockchain topology helps in tracing the changes that occurred in IoT devices [13].
- Credentials:
  All the IoT devices have unique credentials so that the backtracking of every IoT device is possible with the help of blockchain methodology.

Because of the shift from centralized to distributed generation, blockchain technology has emerged as a promising solution for establishing a dependable consensus mechanism. Blockchain allows P2P electronic payments and is categorized into two classes, namely public chains and generalized private chains. The public chain allows every individual or institution to join and leave freely and is a fully open network. A private chain is a network in which participants are subject to restrictions. Blockchain's key characteristics include its decentralized structure, distributed data storage mechanism, consensus agreement, and asymmetric encryption for network security.

2.5.1.1 Blockchain technology infrastructure features

The backbone of infrastructure includes distributed data storage and verification platform based on blockchain, a sensing environment based on IoT, and a service delivery system that is cloud-based [14]. The above cyberinfrastructure supports physical energy entities for generation, transmission, and DNs. In physical infrastructure energy flow, the segment includes generation from power plants and distributed generation resources. Both renewable and nonrenewable power plants are included in the generation segment and their transmission at different levels is considered. Energy entities like energy market systems, load aggregators, and energy retailers are also involved in physical infrastructure.

The cyberinfrastructure has to support the entities being monitored, the need to control equipment, and its coordination with physical infrastructure. In the cyberinfrastructure, the IoT devices are equipped with protocol stacks and communication interfaces to communicate with one another and their users. The blockchain and cloud infra will collect data from multi-domain and stakeholders' decisions are made. The decentralized and distributed processing environments [15] are provided to interface with the IoT so that it provides services like data storage, a platform for smart contracts, and also authorization for agreements and assets. Depending on the application and its requirements, the data collected by IoT can be stored in blockchain or uploaded to the cloud.

2.5.1.2 Application domains of blockchain technology

The application of blockchain technology in future energy systems has huge potential. In a data management system, a blockchain structure can support grid data protection and smart meter data aggregation. Blockchain technology has the potential to reconstruct the energy market so that a centralized trading structure can be adopted. For the above applications, the limitations are cyber-attack vulnerability, difficulty in establishing an open, cross-border energy market system [16], and the fear of a centralized market structure. The blockchain also provides a data synchronization mechanism and an information recording mechanism. The technology also provides authentication services for its energy stakeholders.

2.5.1.3 Challenges of blockchain technology

Though blockchain technology has the potential to provide a solution for future energy systems when applied to practical cases, certain challenges have to be overcome, and a few are highlighted here. Information redundancy occurs as multiple data copies are created on networked nodes, and, also, it would take extra storage space and more power consumption [17]. A performance scalability problem exists in the blockchain system where the consensus protocol can slow down the system when nodes are added. The design flaws and bugs that exist in programs written by a human cannot be fixed or reversed as blockchain techniques are irreversible. Verification tools are to be designed for smart contracts or user-defined programs. Computer network traffic occurs when the blockchain network is coordinated with other infrastructures. Therefore, the differential functional energy organizations have to co-operate for its implementation. The system may suffer due to congestion in the network, deviation in voltage levels, and overloading when integrated with physical grids.

2.5.2 Cloud Computing

Cloud computation systems have developed in recent decades, and the processing of information gathered from systems is the major strategy followed in cloud computing. It mainly offers a solution for the storage, application, and computation of data online, which flows from IoT products. The cloud embodies the "Internet" and computing embodies the computation and processing solutions provided by this strategy [18]. Cloud computing encompasses both Internet application program solutions and data center based hardware systems [19]. Based on the characteristics, fundamental data and structures of multifaceted computation capabilities are processed in cloud computation [20]. The major advantages of using cloud computing systems are that (1) hardware prices could be lowered; (2) computational power and storage ability could be improved; and (3) ease of operation in managing information due to multi-core architectural systems. Additionally, a cloud computational system is a protected system that offers information, computation energy, and storage space that is required from a terrestrial location [21]. Figure 2.5 shows the structure of the IoT in the cloud, and, here, the function of cloud computing is to allow the fundamental data from IoT components to be quickly examined, managed, and sorted resourcefully. Moreover, cloud computing eliminates the costs associated with purchasing software and hardware and operating the algorithms for processing the IoT information, significantly reducing the amount of electrical energy required for neighborhood detail computation.

It is not easy to make each item and tool within the smart grid component of the IoT system and, after that, make all options manageable, controllable, and accessible through cloud computing. Simply said, several issues have

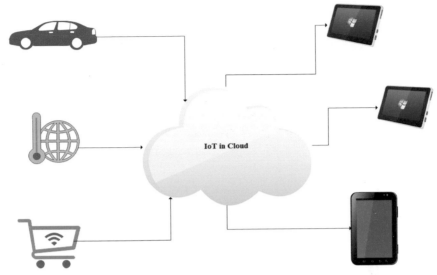

Figure 2.5   Structure of IoT in cloud.

been identified as requiring access to the cloud-IoT to improve the usage of power DNs [22]. The current situations demand that we resolve the cloud computation system from the perspective of the energy industry apart from resources and data. Though the cloud-IoT generates greater opportunities economically in the energy industry, it raises the risk in the system also. Discretion, safety measures, and, predominantly, individuality protection are becoming important aspects within the recurrent and extremely shifting cloud environments, which would be the foundation of public and private clouds integrated with undertakings by companies. In cloud-IoT, diversified networks and solutions might be required and unified to allow for different services and diverse data. The system should have the versatility and adaptability to help all data types based on the needs of program quality support. It can be declared that although the blend of IoT as well as cloud companions might get over several of the present issues of these two solutions, which should be solved distinctly together, brand new tasks might be produced in such a synergistic and joint connection that the following portion intends to talk about several of these difficulties [23].

2.5.2.1 Reference architecture
The IoT plus cloud area is highly heterogeneous, and there are no appropriate requirements for services and data. The standing options have fundamental

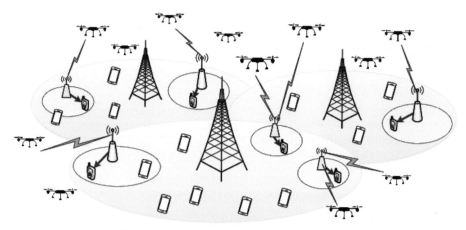

Figure 2.6   Reference architecture of cloud-IoT.

differences and varied ranges concerning the union of IoT and cloud computing systems. This difference is, to a significant degree, because of the absence of regular tools to allow for the style of these remedies, and that deteriorates the intricacy and also necessitates major work for their design and architecture. This issue is aggravated by the absence of proper direction and framework to meet non-functional and functional requirements as these results hinge on their deployment and implementation. Additionally, data produced by various IoT products does not obey any normal format, meaning they are frequently signified in formats that are different, different devices, and so on. Presently, you will find pervasive platforms and no tools for describing IoT equipment and their distinctive capabilities in a uniform and a regular means; so the program agents cannot just do the tasks of theirs, for instance, automated detection and control of products (services, data, and orchestration) [24]. One of several solutions to ease these problems is certainly the reference architecture. The reference architectures can have fun with an important part of the meaning of the structure blocks needed to create architectural methods that mix IoT as well as cloud computing principles. Referring architectures could be observed as intellectual architectures integrating experience and science in a specific program domain, therefore enabling the expansion, interoperability, tuning, and progress of program systems within the specified domain.

Therefore, the recommendations indicated by reference architecture may be viewed as important components in creating cloud-IoT besides their increasing complexity and size. Additionally, because of the benefits of compatibility in acquiring results, this interoperability could be looked after by reformation and producing architectural remedies that are based upon a reference architecture. Among the readily accessible remedies, the open-IoT wedge, as an innovator, appears to be on the road to being realized as the guide structure to combine IoT as well as cloud computing systems by presenting comprehensive explanations and procedures of comprehensive devices [25].

2.5.2.2 Network communication and its challenge

Smart grid equipment and parts that are connected to the Internet generate and send out a vast amount of data that must be categorized, sorted, and enhanced in the cloud. If every piece of information produced by all machines were delivered to the cloud, a substantial stream of information would flow throughout the networks, causing network congestion. As a result, it leads to communication delays, which could be particularly unbearable for apps that are running in real time (or have quasi-real-time performance). Although current cloud systems appear to be capable of providing multiple sources for significant data analytics, widespread usage of these materials can be prohibitively expensive. As a result, it is critical to develop innovative approaches to limit the number of cloud methods used in IoT programs as well as how cloud resources should perform effective processing on the data generated by-products and how to successfully engage computing resources. Cloud computing technology, which enables cloud computing on the system's borders to offer different solutions like processing and analyzing storage space facilities, as well as community services between the platform and connected devices on the cloud, is one of the suggested techniques for dealing with this challenge. A cloud computing system is a more advanced version of traditional cloud computing designed to assist IoT applications with response time constraints, geographical distribution, and mobility requirements. Smart gateways and regional clouds would be required for the notion to be implemented, as they are likely to be used for communication interceding between IoT systems and cloud sources as well as pre-processing and short-term storage of data on smart grids [26].

2.5.2.3 Privacy and security

The adoption of IoT infrastructure on impending smart grids faces significant security challenges. Figure 2.7 shows the security challenges in the IoT. Although security is a huge concern on the current Internet platform, when IoT and cloud computing are combined in smart grids, it becomes a

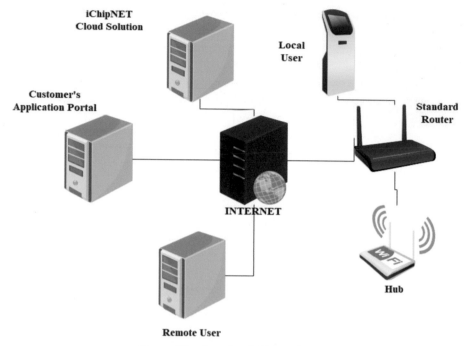

Figure 2.7 Security challenges in IoT.

much larger and more vulnerable issue that may be considered the biggest challenge in IoT. This assumption is supported by the outstanding community redistribution and, as a result, additional access points in the process, as well as the inescapable importance of reliable power service. Furthermore, the items are hooked up to have a simpler design than computer systems, which makes it hard to implement security equipment online ordinarily. The IoT is a great deal closer to actual existence than today's Internet, and the private and important info of customers and energy manufacturers is discussed on the information system. Assaulting such a system will be the same as interfering with the day-to-day operations of smart grid owners in the distribution, transmission, and generation sectors.

The network and protocol security provide an Internet-protected communication mechanism with identity management for authorization and authentication to ensure that information is produced by a specific provider, as well as limiting access, maintaining privacy, and establishing trust between subscribers and entities to aid political aims. Finally, the most serious security problems are IoT attack types such as denial of service (DoS), physical damage, eavesdropping, and node capturing for data extraction by control organizations [27].

2.5.2.4 Background information

Not only is the most advanced sensor data managed delicately in the IoT, but the context of that data, which includes some type of information that could be expected to separate an individual's identity, location, or the state of an item, also necessitates a complex administration process. Contextual elements such as metrics, component status, location, and information precision could be translated into raw data from sensor devices for further processing or retrieval. Even though adding contextual information increases the value of sensed data, it also increases the variability of the data in terms of the format utilized. Combining several screen models creates a barrier in the information sharing and retrieval process. To address the problem, data generated by a variety of IoT-enabled devices must be collected and refined to raise awareness, and access to and authorization for investigation and extraction should be granted to a wide range of end-users and applications [28].

2.5.2.5 Big data analytics

The simple fact that 50 billion products will be connected to the system by 2020 needs a special focus on the conveyance, access, storage, and processing of the massive amounts of data that these devices will generate. IoT technological innovation will undoubtedly be on the list of major components of serious data analytics because of recent scientific growth, and the cloud permits complex analytics and long-term storage. The widespread use of mobile devices and receptors is, in fact, a signal and a request for a scalable computing wedge (approximately 2.5 quadrillion bytes of data is produced every day). Handling this volume of data is a significant challenge, as the program's overall operation closely resembles the functions of the information management services. As an example, cloud-based systems are developing and improving ways to summarize important details based on semantic feature selection algorithms. As a result, after the NoSQL advances, both proprietary and open-source database technology therapies are used to tackle big data [29].

2.5.2.6 Provision of program quality

As the number of datasets grows and the range of unpredictability, uncertainty, and sorts of details in the product expands, the quality of support becomes a contentious issue. Any amount or type of information could be used at any time. Some of the information can also be classified as critical information. As a result, queries must be dynamically prioritized on the cloud side. The bandwidth, noise, delay time, and program missing ratio are commonly used to calculate service quality. The level of service quality must be determined by the data type and the urgency with which the information must be delivered on the sync node (access point) [30]. In protocol supporting various protocols for joining various objects on the Internet, even in the case, they have homogeneous nature are found. For instance, an IoT sensor could use unique protocols, like Wireless HART, IEEE 1451, Zig Bee, and also 6LOW-PAN [31]. The gateway Hojo motor magnetic generator is required for a few protocols, while others cannot be supported. The kind and gate of the sensor utilized decide this. A sensor that is easier or cheaper to obtain has a higher priority than others from the subscriber's and user's perspective. As a result, there is no way to guarantee that the newly added sensor will be profitable. The chart of typical gateway protocols could be used to solve this problem.

2.5.2.7 IPv4 addressing limit

The web is a significant element of the IoT, though the Internet address is restricted to IPv4. Furthermore, numerous solutions, such as CoAP, enable interoperability with embedded devices and elements from the Internet, and new product development activities highlight the necessity to eliminate network address translation (NAT) systems and address these issues with a unique IP address. Simply put, when items have become a part of the Internet (IoT), they need a unique and exclusive identification, and Figure 2.8 shows the interface in the industrial platform. Furthermore, mobile items, such as moveable sensor nodes on all types of cars (especially electric

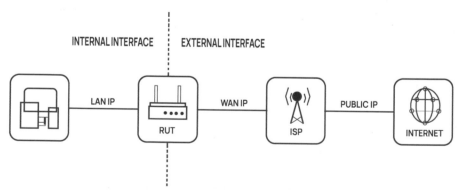

Figure 2.8   Interface in industrial platform.

vehicles) and other moving objects, must map their real-time identification to the new community they have entered. Assigning IPv6 addresses to this pervasive network is wise since the IPv6 address space is deemed to be ample to handle IoT systems. To overcome this issue, IPv6 employs the IP128 short address, which offers benefits such as Internet integration, continuous connectivity, and rest compliance. SLoWPAN and Zig Bee IP protocols have good compatibility with IPv6 in IoT-enabled embedded goods.

However, there are currently very few industrial platforms that have these capabilities. Because of IoT adoption, devices will transition from human-centered pursuits to networks featuring machine-to-human and machine-to-machine connections, and IPv6 will pave the way. If IPv6 is used to recognize the communications of objects, the widespread deployment of IPv6 becomes a contentious issue. Unless a proper, regular, and effective system for cohabitation of IPv6 and IPv4 is implemented, the IPv6 to tools (objects) project will not be successful [32]. According to research focused on the coexistence of IPv6 and IPv4, the smooth transition to IPv6 should be further developed.

2.5.2.8 Legal aspects and social facts

These two critical concerns are linked to one another in some way. Consider a cloud-based, data-driven service provided by a single person. On the one hand, the main service provider must adhere to various international regulations, while, on the other hand, consumers must be motivated to participate in data collection. It is easier to provide people with the option of collaborating on information-gathering items that represent a specific concept. Recently, several issues have been identified that may be found in almost any system, such as the use of people as sensor information sources. The response also contributes to the users' continued loyalty (such as combining human-made qualitative observations with quantitative observations produced by machines; or maybe they want to determine and control reliability, reputation, data quality, and belief in) [33]. Subscribers and users may be enabled by up-to-date customized applications as well as blocks such as toolkits, frameworks, and accelerators that allow users to engage with IoT in the same way that they would with the Internet. Such gadgets must allow design experts to learn about consumer performance while also providing owners with an engaged role in the engineering design system. As a result, to achieve this goal, the resources should make it possible for specific people to see different design characteristics simply and cost-effectively. To address this problem, scientists and experts are working to develop appropriate tools for using a collaborative sampling technique in which both designers and specific users uncover applications and connections.

2.5.2.9 Service detection

The cloud supervisor, or possibly agent in cloud-IoT, oversees the identification of new services and goods for individual consumers. On the IoT, any item can become a component of the system at any time, and the system can become a component at any time. Some IoT nodes are capable of mobility. Finding new solutions, as well as their instantaneous status and updating the currently established services, remains a contentious issue. For a larger and more complicated IoT, management equipment may be necessary, which is responsible for monitoring the condition of IoT nodes, keeping an eye on cellular nodes and maintaining the current state of both the present IoT and recently added IoT nodes. For this work, a consistent means of obtaining assistance is required [34].

2.6 Conclusion

With the rise in global demand for energy and the numerous barriers to its extraction, like fossil fuel depletion, attention toward climate change problems, etc., the world is facing a new shift in energy generation. The advent of IoT and its smart integration with current power distribution systems have tremendously played a role in reducing the socioeconomic-environmental impacts of the current energy distribution system, thus fulfilling the needs of energy use. Faced with several challenges in this type of integrated power distribution system, the state-of-the-art IoT power sector is poised to move from the latest hierarchy to build a smarter, decentralized system. In this study, we discuss the role of the IoT in the energy supply chain and the application of smart grids in the power distribution system. Furthermore, this article discusses the role of the IoT and its structure and challenges in integrating it into the energy chain as well as tackling these challenges with IoT support and its benefits. This study reviews the many components of an IoT system, such as supporting communication and sensor technologies, and how they might be used in the energy sector, such as temperature, humidity, light, speed, passive infrared, and proximity sensors. Also, this study discusses cloud computing and data analytic platforms, which are data analysis and visualization tools that may be used for a variety of smart energy applications, from buildings to smart cities. Moreover, this chapter highlights the after-implementation challenges of IoT in energy management systems, such as privacy problems and challenges, security problems and their challenges, and data storage problems and their challenges. This chapter discusses the IoT challenges and their solutions through blockchain technology and cloud computing briefly like the future scope of the research.

References

[1] Friess, D. A. (2016). Ecosystem services and disservices of Mangrove forests: Insights from historical colonial observations. *Forests*, 7(9), 183. https://doi.org/10.3390/f7090183.

[2] Gardoni, P., & Murphy, C. (2020). Society-based design: Promoting societal well-being by designing sustainable and resilient infrastructure. *Sustainable and Resilient Infrastructure*, 5(1-2), 4–19.

[3] Li, Z. Q., Xu, H., & Zhang, Y. (2014). Remote sensing of haze pollution based on satellite data: Method and system design. *Environmental Monitoring in China*, 30(03), 159–165.

[4] Parker, D. S. (2020). The implementation of the Internet of Things (IoT): A case study of the barriers that prevent implementation of IoT within small and medium enterprises (SME), Doctoral dissertation, Northcentral University.

[5] Muruganantham, B., Gnanadass, R., & Padhy, N. P. (2017). Challenges with renewable energy sources and storage in practical distribution systems. *Renewable and Sustainable Energy Reviews*, 73, 125–134.

[6] Sinsel, S. R., Riemke, R. L., & Hoffmann, V. H. (2020). Challenges and solution technologies for the integration of variable renewable energy sources—A review. *Renewable Energy*, 145, 2271–2285.

[7] Bhattacharjee, S., & Nandi, C. (2019). Implementation of industrial internet of things in the renewable energy sector. In *The Internet of Things in the Industrial Sector*, pp. 223–259, Springer, Cham.

[8] Serpanos, D., & Wolf, M. (2018). Industrial Internet of Things. In *Internet-of-Things (IoT) Systems*, pp. 37–54, Springer, Cham.

[9] Al-Ali, A. R., Zualkernan, I. A., Rashid, M., Gupta, R., & AliKarar, M. (2017). A smart home energy management system using IoT and big data analytics approach. *IEEE Transactions on Consumer Electronics*, 63(4), 426–434.

[10] Salkuti, S. R. (2020). Challenges, issues and opportunities for the development of smart grid. *International Journal of Electrical and Computer Engineering (IJECE)*, 10, 1179–1186.

[11] Pilkington, M., (2016). Blockchain technology: Principles and applications. In *Research Handbook on Digital Transformations*, vol. 225, Edward Elgar Publishing, Cheltenham, U.K.

[12] .Veena, P., Panikkar, S., Nair, S. & Brody, P. (2015). Empowering the edge-practical insights on a decentralized Internet of Things. *Empowering Edge Practical Insights: A Decentralized Internet Things*, 17, 2015.

[13] Prisco, G. (Nov. 2015). Slock.it to introduce smart locks linked to smart Ethereum contracts decentralize the sharing economy. [Online] Available: https://bitcoinmagazine.com/articles/sloc-it-tointroduce-smart-locs-lined-to-smart-ethereumcontractsdecentralizethe-sharing-econ omy-1446746719.

[14] Stergiou, C., Psannis, K. E., Kim, B. G., & Gupta, B. (2018). Secure integration of IoT and cloud computing. *Future Generation Computer Systems*, 78, 964-975.

[15] Ji, C., Li, Y., Qiu, W., Awada, U., & Li, K. (Dec. 13–15, 2012). Big data processing in cloud computing environments. In *Proceedings of the 2012 12th International Symposium on Pervasive Systems, Algorithms and Networks*, San Marcos, TX, USA, pp. 17–23, vol. 107.

[16] Vukolic, M. (2015). The quest for scalable blockchain fabric: Proof-of-work vs. BFT re-plication. In *Proceedings of the International Workshop on Open Problems in Network Security*, Springer, Berlin, pp. 112–125.

[17] Blockchain. (2017). Transactions per block. [Accessed 2-Nov-2017]. Available: https://blockchain.info/charts/n-transactions-per-block

[18] Vytelingum, P., Ramchurn, S. D., Voice, T. D., Rogers, A., & Jennings, N. R. (2010). Trading agents for the smart electricity grid. In *Proceedings of the 9th International Conference on Autonomous Agents and Multi-Agent Systems (IFAAMAS)*, pp. 897–904.

[19] Block, C., Neumann, D., & Weinhardt, C. (2008). A market mechanism for energy allocation in micro-CHP grids. In *Proceedings of the 41st Annual Hawaii International Conference on System Sciences*, IEEE, p. 172.

[20] Block, C. A., Collins J., Ketter W., & Weinhardt, C. (2009). A multi-agent energy trading com- petition. [Accessed 22-Nov-2017]. Available: www.erim.eur.nl

[21] Ketter, W., Collins, J., & Reddy, P. (2013). Power TAC: A competitive economic simulation of the smart grid. *Energy Economics*, 39, 262–270.

[22] Yassine, A., Singh, S., Hossain, M. S., & Muhammad, G. (2019). IoT big data analytics for smart homes with fog and cloud computing. *Future Generation Computer Systems*, 91, 563–573.

[23] Carvalho, O., Garcia, M., Roloff, E., Carreño, E. D., & Navaux, P. O. (Sep. 2017). IoT workload distribution impact between edge and cloud computing in a smart grid application. In *Proceedings of the Latin American High Performance Computing Conference*, Springer, Cham, pp. 203–217.

[24] Forcan, M., & Maksimoviæ, M. (2019). Cloud-fog-based approach for smart grid monitoring simulation. *Modelling Practice and Theory*, 101988.

[25] Jamborsalamati, P., Fernandez, E., Hossain, M. J., & Rafi, F. H. M. (Nov. 2017). Design and implementation of a cloud-based IoT platform for data acquisition and device supply management in smart buildings. In *Proceedings of the 2017 Australasian Universities Power Engineering Conference (AUPEC)*, IEEE, pp. 1–6.

[26] Stergiou, C., Psannis, K. E., Kim, B. G., & Gupta, B. (2018). Secure integration of IoT and cloud computing. *Future Generation Computer Systems*, 78, 964–975.

[27] Chiang, M., & Zhang, T. (2016). Fog and IoT: An overview of research opportunities. *IEEE Internet of Things Journal*, 3(6), 854–864.

[28] Cai, H., Xu, B., Jiang, L., & Vasilakos, A. V. (2016). IoT-based big data storage systems in cloud computing: perspectives and challenges. *IEEE Internet of Things Journal*, 4(1), 75–87.

[29] Zahoor, S., Javaid, S., Javaid, N., Ashraf, M., Ishmanov, F., & Afzal, M. (2018). Cloud–fog–based smart grid model for efficient resource management. *Sustainability*, 10(6), 2079.

[30] Mekala, M. S., & Viswanathan, P. (Aug. 2017). A survey: smart agriculture IoT with cloud computing. In *Proceedings of the 2017 international conference on microelectronic devices, circuits and systems (ICMDCS)*, IEEE, pp. 1–7.

[31] Aazam, M., Huh, E. N., St-Hilaire, M., Lung, C. H., & Lambadaris, I. (2016). Cloud of things: Integration of IoT with cloud computing. In *Proceedings of the Robots and Sensor Clouds*, Springer, Cham, pp. 77–94. [45]. Stojmenovic, I. (Nov. 2014). Fog computing: A cloud to the ground support for smart things and machine-to-machine networks. In *Proceedings of the 2014 Australasian Telecommunication Networks and Applications Conference (ATNAC)*, IEEE, pp. 117–122.

[32] Kaur, M. J., & Maheshwari, P. (Mar. 2016). Building smart cities applications using IoT and cloud-based architectures. In *Proceedings of the 2016 International Conference on Industrial Informatics and Computer Systems (CIICS)*, IEEE, pp. 1–5.

[33] Tom, R. J., & Sankaranarayanan, S. (Jun. 2017). IoT based SCADA integrated with fog for power distribution automation. In *Proceedings of the 2017 12th Iberian Conference on Information Systems and Technologies (CISTI)*, IEEE, pp. 1–4.

[34] Khan, Z., Anjum, A., & Kiani, S. L. (Dec. 9–12, 2013). Cloud based big data analytics for smart future cities. In *Proceedings of the 2013 IEEE/ACM 6th International Conference on Utility and Cloud Computing*, Dresden, Germany, pp. 381–386.

3

Power Quality Monitoring of Low Voltage Distribution System Toward Smart Distribution Grid Through IoT

P. Sivaraman[1], C. Sharmeela[2], S. Balaji[3], P. Sanjeevikumar[4], and S. Elango[5]

[1]Vestas Technology R&D Chennai Pvt Ltd, India
[2]Anna University, India
[3]IIT Kanpur, India
[4]Aarhus University, Denmark
[5]Coimbatore Institute of Technology, India
E-mail: sivaramanp@ieee.org; sharmeela20@yahoo.com;
sanjeevi_12@yahoo.co.in

Abstract

Smart distribution grid allows bidirectional power flow in the distribution system with a mixture of multiple renewable energy sources. The distribution system is highly affected by various power quality (PQ) disturbances like harmonics, transients, undervoltage, overvoltage, unbalance, etc. Ensuring the quality of power is supplied to the consumers is one of the major concerns of the distribution company. The distribution company has to monitor the PQ parameters at various locations of the smart distribution grid. This chapter discusses the monitoring of various PQ problems, such as undervoltage, overvoltage, interruption, and overload of a smart distribution system. It employs the remote communication stations (RCS) at the various locations of a smart distribution system. RCS monitors the various PQ parameters continuously and compares them with the reference value. It gives notifications whenever the PQ parameters exceed the reference values. This Internet of Things based monitoring system detects the PQ events such as undervoltage and overvoltage, unbalanced voltage, and overload. It sends the details to the distribution company/users using alert SMS and also on the web.

Keywords: Internet of Things (IoT), smart grid, distribution system, power quality (PQ), GSM.

3.1 Introduction

The distribution system is the last part of the power system, distributing electric power to the end-users. The distribution system is the electrical system between the sub-station and the consumer's meters fed by the distribution system [1]. It comprises feeders, distributors, and service mains. An AC distribution system is classified into a primary distribution system and secondary distribution system. The primary distribution system voltages vary in different countries. The typical primary distribution voltages are 33, 22, 11, kV carried by 3Φ, 3-W system and secondary distribution system employs 415 V carried by 3Φ, 4-W system [2–5, 26]. Distribution systems comprise residential, commercial, and industrial systems, and it distributes the electric power supply to the end-users [6–8]. The distribution system shall have high supply reliability, efficiency, and power supply quality to its end-users [3]. To have such a distribution system, certain areas need to be monitored and controlled. They are, location of fault identification, clearing the fault in lesser time, effectively monitoring the power quality (PQ; voltage and current variation), maintaining the voltage drop and PQ variation within the acceptable tolerance limits, and maintaining the proper database of failure/outage events [9, 10].

The distribution system is highly affected by various PQ disturbances, such as harmonics, transients, undervoltage, overvoltage, unbalance, interruption, flicker, etc. [11–14]. The poor PQ results in premature equipment failure, mal-operations, tripping of sensitive critical equipment, and financial losses to end-users as well as distribution companies [15, 22]. Hence, it is essential to monitor the PQ parameters at various locations of the distribution grid to find out quality of the power supply is delivered to the end-users.

The equipment shall monitor the PQ parameters, such as voltage and current transformer, PQ meter, or analyzer. With this equipment, PQ parameters like voltage sag, swell, transients, harmonics, flicker, etc., shall be monitored/measured at the measurement point [16, 17]. There are two types of PQ measurement carried out at the site. The first one is doing the measurement after the failure of any devices/equipment, i.e., finding the reason why the device/equipment failed. This method is widely used to find the reason for failure and then propose a mitigation plan to avoid equipment failure in the future. The second one is doing the measurement periodically or regularly as preventive maintenance to identify any existing PQ problem and mitigate it before it leads to equipment failure, mal-operation, etc. Either by permanent PQ monitoring (by means of permanent current transformer, voltage transformer, and PQ analyzer at the site installation) or temporary PQ monitoring through portable equipment (portable current transformer, voltage transformer, and PQ analyzer at the site installation). In permanent-type monitoring, the capital investment cost is high. Hence, in many places, portable PQ instruments are used for temporary PQ monitoring.

In recent years, the Internet of Things (IoT) has been widely used to monitor and control equipment from a remote location or centralized location [21]. It acts as a mediator between the physical equipment (hardware) and application (software). IoT-based technology is used to monitor the PQ parameters from the remote location or centralized location and the actual measurement site as well. It is using various network technologies (wireless communication technologies) as a communication platform for remote monitoring [18–20, 23–25]. This chapter discusses IoT-based PQ monitoring in a smart distribution system.

3.2 Introduction to Various PQ Characteristics

As per IEEE Std 1100-2005, the PQ is defined as powering and grounding electronic equipment in a manner suitable to the operation of that equipment and compatible with the premise wiring system and other connected equipment [9]. The PQ characteristics as per IEEE Std 1159-2019 are as follows [10]:

- short duration root mean square (RMS) variation;
- long duration RMS variation;
- transients;
- unbalance or imbalance;
- voltage fluctuations;
- waveform distortion;
- power frequency variations.

Transients are momentary changes/variations in voltage or current or both due to lightning, faults, switching operations, etc.

Short duration RMS variation is called a variation in RMS voltage for between 0.5 cycles and 60 seconds.

Long duration RMS variation is called a variation in RMS voltage of over 60 seconds.

Imbalance or unbalance is the ratio of the magnitude of the negative sequence component to the positive sequence component expressed in percentage in a three-phase system.

Voltage fluctuation is called a continuous change in the voltage (instantaneous cycle-to-cycle voltage) because of connected load resistance change in every cycle.

Undervoltage is called long-duration RMS voltage magnitude reduction between 0.9 and 0.8 pu for over 1 minute. The causes of undervoltages are continuous energization of large capacity loads, overloaded circuits, switching OFF capacitor banks, and undervoltage.

Overvoltage is called long-duration RMS voltage magnitude that increases above 1.1 pu for over 1 minute. The causes of overvoltage are switching OFF large capacity loads, switching ON capacitor bank, and incorrect tap operation of transformers.

Sustained interruptions are called long-duration RMS voltage magnitude reduction of less than 0.1 pu for over 1 minute.

Waveform distortion is called steady-state distortion or deviation from the ideal sinusoidal characteristics [11].

3.3 Introduction to IoT

IoT is a medium of communication or mediator between hardware equipment and software. The typical structure of IoT-based monitoring is shown in Figure 3.1. In this method, all the metering and controlling equipment are connected to a centralized or common platform, i.e., IoT. It will collect the data from the connected hardware equipment and transfer it to the customer application's storage and data analytics.

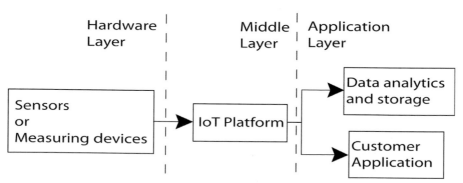

Figure 3.1 Typical structure of IoT-based monitoring.

It comprises three layers, namely hardware layer (sensors or measuring equipment), middle layer (IoT platform), and application layer (customer application, data analytics, and storage) [29, 30]. The hardware layer comprises measuring types of equipment like a current transformer, voltage transformer, temperature sensor, humidity sensor, infrared sensor, photodetectors, gas sensors, pressure sensors, ultrasonic sensors, etc. The middle layer receives the signals from the hardware layer and transfers these signals to the application layer through communication mediums like GSM, GPRS, etc. Finally, the application layer receives the input from the middle layer, analyzes the received input (comparing the value with pre-defined/reference value), stores it, and communicates to the users using human-machine interface (HMI).

3.4 Smart Monitoring using IoT for the Low Voltage Distribution System

Electricity is one of the basic needs of people in the modern world. It is very important to have quality and reliability of supply distributed to the consumer's equipment in the distribution system. To maintain the quality and reliability of the supply, the following areas should be improved in a distribution system:

- identifying the faults that occurred at the instant;
- clearing the fault in lesser time;
- maintaining the proper database of fault and outage event;
- monitoring the quality of power;
- maintaining the voltage drop and PQ variation within the acceptable tolerance limits.

The typical block diagram of IoT-based monitoring of low voltage distribution systems is shown in Figure 3.2.

It comprises voltage and current transformers/sensors, microcontroller, and communication medium (transmitter) at the remote site location where the actual monitoring has been carried out. The receiver receives the input signal from the remote site location at the central monitoring and control center, processes and analyzes it, stores it, and communicates to the users through the monitor, alarm signal, warning signal, etc.

From the literature, many wireless communication technologies are used to communicate/transfer the data between remote sites and central monitoring locations [28]. The following wireless communication technologies are widely used for communication:

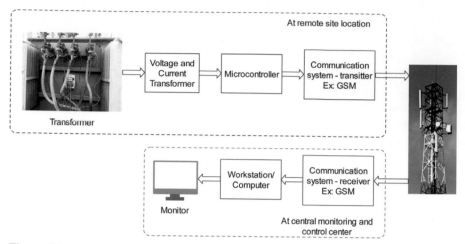

Figure 3.2 Typical block diagram of IoT-based monitoring of low voltage distribution system.

1. Bluetooth communication;
2. Wi-Fi communication;
3. ZigBee communication;
4. GPS communication;
5. GPRS communication;
6. GSM communication.

1. Bluetooth Communication:

The Bluetooth communication technology is mostly used to establish communication between mobile phones for data transfer in a short distance. This method has a limited connectivity coverage area and is insecure, and free from interference [27].

2. Wi-Fi Communication:

Like Bluetooth, Wi-Fi communication technology is also used for connection between multiple equipment at lesser coverage distance. This method is widely used for providing the Internet connection to the devices.

3. ZigBee Communication:

The wireless ZigBee communication is used for networking and connectivity between the devices based on IEEE Std 802.15.4. This method is more suitable for connecting many devices due to its network structure flexibility.

4. GPS Communication:

The global positioning system, in short form GPS, is satellite-based wireless communication technology. This method offers a wider coverage range with an accuracy range of +/-10 m.

5. GPRS Communication:

The general packet radio service, in short form GPRS, is widely used in mobile phone communication across the globe. This method enables the wireless connection from anywhere within the coverage range.

6. GSM Communication:

The global system for mobile, in short form GSM, is a widely used mobile communication method. It has the frequency range of either 900 or 1800 MHz.

3.5 Power Quality Monitoring of Low Voltage Distribution System – Case Study

The single line diagram (SLD) of the circuit is considered for the PQ measurement as shown in Figure 3.3.

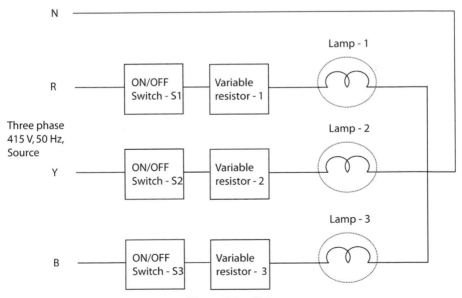

Figure 3.3   SLD.

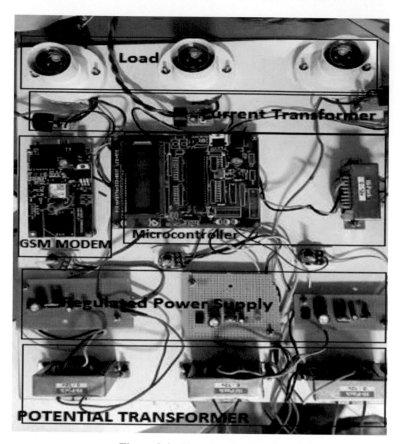

Figure 3.4 Experimental setup.

The experimental setup for the SLD shown in Figure 3.3 with the smart monitoring system is shown in Figure 3.4. It consists of three-phase, 415-V, 50-Hz power source, loads (three numbers of single-phase lamps), microcontroller, GSM modem, current transformer, and potential transformer.

The control block diagram of the experimental setup shown in Figure 3.4 is shown in Figure 3.5. The PIC16F877A receives the load (lamps) voltage and the current using the potential transformer and the current transformer, respectively. It compares the voltages and current magnitudes with respect to reference values (limits). If the limits are violated, an alarm is triggered in addition to a pop-up on the LCD screen. The LCD and buzzer are used to communicate the deviations in the measured voltage and current to the users locally, while GSM-based communication is used for remote monitoring.

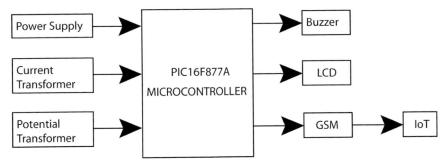

Figure 3.5 Block diagram.

GSM modem can accept any GSM network as a SIM card and just like a mobile phone with its unique phone number. This modem uses the RS232 port for communication with embedded applications. The SIM900A is a complete dual-band GSM/GPRS solution in an SMT module featuring an industry-standard interface; the SIM800 delivers GSM/GPRS 900/1800 MHz performance for voice, SMS, data, and fax in a small form factor and with low power consumption. The LCD can perform the local monitoring; remote monitoring can be done through the website and mobile via SMS.

3.5.1 Undervoltage

An undervoltage is defined as a reduction of RMS voltage magnitude for over a 1-minute time duration below specified limits. The nominal voltage is 240 V with a tolerance of ±10%. Hence, the reference voltage (lower limit) is set as 216 V. If the monitored voltage is less than 216 V for over 1 minute, the microcontroller sets off an alarm using the buzzer in addition to a pop-up message on the LCD as "undervoltage." The flow chart of undervoltage identification is shown in Figure 3.6.

If the consumer's phase voltage reduces below the reference value, an alert notification is sent via SMS using the GSM module. Additionally, the same information is recorded and communicated through a webpage. Figure 3.7 illustrates the notifications received by the customer through SMS text.

3.5.2 Overvoltage

An overvoltage is defined as the increase in RMS voltage magnitude over a 1-minute time duration. The nominal voltage is 240 V with a tolerance of ±10%. Hence, the reference voltage is set as 264 V. If the measured

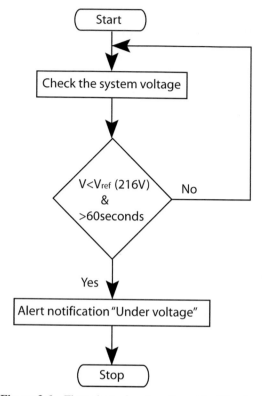

Figure 3.6 Flow chart of undervoltage identification.

Figure 3.7 SMS alert for undervoltage.

voltage is higher than 264 V for over 1 minute, then the microcontroller sets off an alarm using the buzzer in addition to a pop-up message on the LCD as "overvoltage." The flow chart of overvoltage identification is shown in Figure 3.8.

Figure 3.9 illustrates the SMS received by the consumer during the advent of an upper limit violation on the voltage.

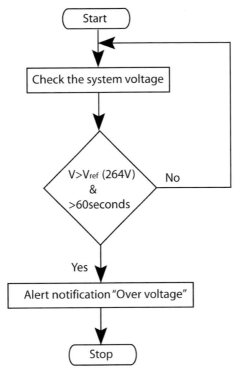

Figure 3.8 Flow chart of overvoltage identification.

Figure 3.9 SMS alert for overvoltage.

3.5.3 Interruption

An interruption is defined as a decrease in RMS voltage magnitude less than 0.1 pu. If the time duration is less than 1 minute, then it is called a momentary interruption, and if it is 1 minute, it is called a sustained interruption. The nominal voltage is 240 V with a tolerance of $\pm10\%$. Hence, the reference voltage is set as 24 V. If the measured voltage is lesser than 24 V for less than 1 minute which is a momentary interruption and over 1-minute time duration which is sustained interruption, then the microcontroller will send an alarm through the buzzer and pop-up through LCD as an interruption. The flow chart of interruption identification is shown in Figure 3.10.

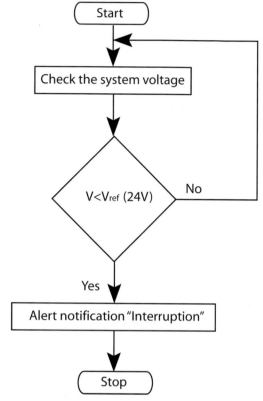

Figure 3.10 Flow chart of interruption identification.

Figure 3.11 SMS alert for voltage interruption.

The end-user will receive the SMS in mobile phone for voltage interruption alert as shown in Figure 3.11.

3.5.4 Overload in Branch Circuit

An overload may be termed as the increase in RMS current drawn by the load exceeding its nominal value. As a result of the higher load current, the current flow in the branch circuits is increased. This increase in current is known as overloading of branch circuits. For example, the nominal current of the load is 100 A. If the current flow in this branch is over 100 A, it is called overloading. Up to 10% of the overloading (on the nominal value) is acceptable. To identify the branch circuit's overloading, the reference current is set as 0.2 A. If the measured current is higher than 0.2 A, the microcontroller will send an alarm through the buzzer and pop-up through the LCD as overcurrent. The flow chart of overvoltage identification is shown in Figure 3.12.

The end-user will receive the SMS in mobile phone for overcurrent alert as shown in Figure 3.13.

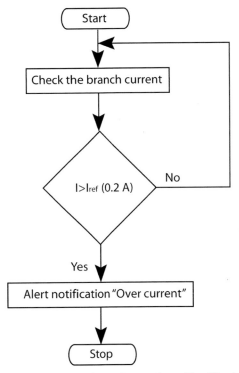

Figure 3.12   Flow chart of overvoltage identification.

Figure 3.13   SMS alert for overvoltage.

S.No	Date	Area	Phase1 V	Phase2 V	Phase3 V	Current1	Current 2	Current 3
1	2018/03/18 05:27:44pm	1	2	3	3	4	5	6
2	2018/03/18 05:28:14pm	1	2	3	3	4	5	6
3	2018/03/18 05:50:27pm	1	0230	0158	0204	0003	0012	0004
4	2018/03/18 06:01:56pm	1	0088V	0112V	0126V	0000A	0007A	0005A
5	2018/03/18 06:02:33pm	1	0186V	0198V	0220V	0014A	0012A	0004A
6	2018/03/18 06:03:50pm	1	0186V	0197V	0221V	0006A	0012A	0004A
7	2018/03/18 06:04:21pm	1	0186V	0197V	0202V	0000A	0011A	0005A
8	2018/03/18 06:04:52pm	1	0186V	0197V	0202V	0000A	0011A	0004A
9	2018/03/18 06:05:27pm	1	186V	198V	202V	000A	012A	006A
10	2018/03/18 06:05:58pm	1	186V	198V	202V	009A	012A	006A
11	2018/03/18 06:06:28pm	1	186V	198V	203V	007A	012A	006A
12	2018/03/18 06:06:59pm	1	186V	198V	203V	001A	012A	006A
13	2018/03/18 06:07:31pm	1	187V	198V	203V	016A	010A	005A

Figure 3.14　Screenshot of real-time monitoring through web.

Through this IoT-based monitoring system, consumers will immediately be notified about the current PQ parameters through the mobile communication system via SMS. This enables supervision/monitoring of the real-time PQ parameters. The real-time web-based monitoring of low voltage smart distribution can be monitored anywhere in the world. The data can be accessed through the web link http://vehiccleiot.in/distribution/view.php. Figure 3.14 enumerates the real-time parameters displayed on the webpage.

3.6 Conclusion

In the modern world, electricity is one of the basic needs of human beings. Supplying the quality and high reliable power supply is the most important concern for all the distribution companies. This chapter discusses the PQ monitoring at a low voltage smart distribution system through continuous monitoring with advanced features like analyzing the PQ parameter, storing/keeping the proper database, etc. It also sends the alert notification to the users via SMS and real-time monitoring on the web page. Power distribution companies can supply a more reliable power supply to their customer by using this monitoring system.

References

[1] P. Sivaraman and C. Sharmeela, Introduction to electric distribution system. In B. Kahn, H. H. Alhelou, and Ghassan (Eds.), *Handbook of Research on New Solutions and Technologies in Electrical Distribution Networks*. Hershey, PA: IGI Global, 2020, pp. 1–31.

[2] P. Sivaraman and C. Sharmeela, Existing issues associated with electric distribution system. In B. Kahn, H. H. Alhelou, and Ghassan (Eds.), *Handbook of Research on New Solutions and Technologies in Electrical Distribution Networks*. Hershey, PA: IGI Global, 2020, pp. 1–31.

[3] T.S. Somkuwar and M.G. Panjwani, Review paper on electrical distribution line monitoring, *International Journal of Advance Research in Computer and Communication Engineering*, 2015.

[4] P. Sivaraman, D. Gunapriya, K. Parthiban, and S. Manimaran, Hybrid fuzzy PSO algorithm for dynamic economic load dispatch, *Journal of Theoretical and Applied Information Technology*, vol. 62, no. 3, pp. 794–799, April 2014.

[5] P. Sivaraman, S. Manimaran, K. Parthiban, and D. Gunapriya, PSO approach for dynamic economic load dispatch problem, *International Journal of Innovative Research in Science, Engineering and Technology*, vol. 3, no. 4, pp. 11905–11910, April 2014.

[6] P. Sivaraman, C. Sharmeela, R. Mahendran, and A. Thaiyal Nayagi, *Basic Electrical and Instrumentation Engineering*. Hoboken, NJ: John Wiley & Sons, 2020.

[7] P. Sivaraman, C. Sharmeela, and D.P. Kothari, Enhancing the voltage profile in distribution system with 40 GW of solar PV rooftop in Indian grid by 2022: A review, In *Proceedings of 1st International Conference on Large Scale Grid Integration of Renewable Energy in India*, September 2017.

[8] IEEE Std 141-1993, IEEE Recommended Practice for Electrical Power Distribution for Industrial Plants.

[9] IEEE Std 1100-2005, IEEE Recommended Practice for Powering and Grounding Electronic Equipment.

[10] IEEE Std 1159-2019, IEEE Recommended Practice for Monitoring Electric Power Quality.

[11] P. Sivaraman and C. Sharmeela, Power quality problems associated with electric vehicle charging infrastructure. In P. Sanjeevikumar, C. Sharmeela, and J. B. Holm-Nielsen (Eds.), *Power Quality in Modern Power Systems*. Amsterdam: Elsevier, 2021.

[12] R. Mahendran, P. Sivaraman, and C. Sharmeela, Three phase grid interfaced renewable energy source using active power filter, In *Proceedings 5th International Exhibition & Conference, GRIDTECH 2015*, New Delhi, India, April 2015, pp. 77–85.

[13] P. Sivaraman and C. Sharmeela, Power quality and its characteristics, In P. Sanjeevikumar, C. Sharmeela, and J.B. Holm-Nielsen (Eds.), *Power Quality in Modern Power Systems*. Cambridge, MA: Academic Press, 2021.

[14] P. Sivaraman and C. Sharmeela, Power system harmonics, In P. Sanjeevikumar, C. Sharmeela, and J.B. Holm-Nielsen (Eds.), *Power Quality in Modern Power Systems*. Cambridge, MA: Academic Press, 2021.

[15] P. Sivaraman and C. Sharmeela, Battery energy storage system addressing the power quality issue in grid connected wind energy conversion system, In *Proceedings of the 1st International Conference on Large-Scale Grid Integration Renewable Energy in India*, New Delhi, India, September 2017.

[16] P. Sanjeevikumar, C. Sharmeela, J.B. Holm-Nielsen, and P. Sivaraman, *Power Quality in Modern Power Systems*. Cambridge, MA: Academic Press, 2021.

[17] C. Sharmeela, P. Sivaraman, P. Sanjeevikumar, and J.B. Holm-Nielsen, *Microgrid Technologies*. Hoboken, NJ: John Wiley & Sons, 2021.

[18] B. Li, B.Z.J. Guo, and H. Yao, Study of cognitive radio based wireless access communications of power line and substation monitoring system of smart grid, In *Proceedings of the International Conference On Computer Science and Service System*, IEEE, 2012.

[19] V.C. Gungor, B. Lu, and G.P. Hancke, Opportunities and challenges of WSN in smart grid, *IEEE Transactions on Industrial Electronics*, vol. 57, no. 10, October 2010.

[20] J. Gutiérrez, J.F. Villa-Medina, A. Nieto-Garibay, and M.. Porta-Gándara, Automated irrigation system using a wireless sensor network and GPRS module, *IEEE Transactions on Instrumentation and Measurement*, vol. 63, no. 1, January 2014.

[21] P. Sivaraman and C. Sharmeela, IoT-based battery management system for hybrid electric vehicle, In A. Chitra, P. Sanjeevikumar, J. B. Holm-Nielsen, and S. Himavathi (Eds.), *Artificial Intelligent Techniques for Electric and Hybrid Electric Vehicles*. Beverly, MA: Scrivener Publishing LLC, 2020, pp. 1-16.

[22] C. Sharmeela, P. Sivaraman, and S. Balaji, Design of hybrid DC mini-grid for educational institution: Case study, *Lecture Notes in Electrical Engineering*, vol. 580, pp. 125-134, 2020.

[23] L. Atzori, A. Iera, and G. Morabito, The Internet of Things: A survey, *Computer Networks*, vol. 54, pp. 2787–2805, 2010.

[24] N. Harish, V. Prashal, and D. Sivakumar, IoT based battery management system, *International Journal of Applied Engineering and Research*, vol. 13, pp. 5711–5714, 2018.

[25] S. Balaji and S. Usa, Characterization of polypropylene under DC-polarity reversal, In *Proceedings of the 2019 IEEE 4th International Conference on Condition Assessment Techniques in Electrical Systems (CATCON)*, 2019, pp. 1-5, doi: 10.1109/CATCON47128.2019.CN0064.

[26] N. Sharma, M. Shamkuwar, and I. Singh, The history, present, and future with IoT, *Intelligent Systems Reference Library*, pp. 27–51, 2018.

[27] P. Sivaraman and C. Sharmeela, Solar micro-inverter, In J. Zbitou, C. Pruncu, and A. Errkik (Eds.), *Handbook of Research on Recent Developments in Electrical and Mechanical Engineering*. Hershey, PA: IGI Global, 2020, pp. 283–303.

[28] M. Weiser, The computer for the 21st century, *ACM SIGMOBILE Mobile Computing and Communications Review*, vol. 3, no. 3, pp. 3–11, 1999.

[29] M. Ersue, D. Romascanu, J. Schnwlder, and A. Sehgal, Management of networks with constrained devices: Use cases, no. 7548. RFC Editor, May 2015.

[30] S. Balaji, V.S. Devendran, and S. Chenniappan, Electromagnetic analysis of underground cables for upcoming smart city: Case study, In *Proceedings of the 2020 IEEE International Conference on Power and Energy (PECon)*, 2020, pp. 259-263, doi: 10.1109/PECon48942.2020.9314306.

[31] T. de Vass, H. Shee, and S. Miah, IoT in supply chain management. Opportunities and challenges for businesses in early Industry 4.0 context, *Operations and Supply Chain Management: An International Journal*, pp. 148–161, 2021.

[32] H. Tang, S. Lu, G. Qian, J. Ding, Y. Liu, and Q. Wang, IoT-based signal enhancement and compression method for efficient motor bearing fault diagnosis, *IEEE Sensors Journal*, vol. 21, pp. 1820-1828, 2021.

[33] S. Balaji and T. Ayush, Placement of energy exchange centers and bidding strategies for smartgrid environment, 2020 Wiley, Smart Grid Technologies. ISBN: 978-1-119-71087-5.

4

Health Monitoring of a Transformer in a Smart Distribution System using IoT

P. Sivaraman[1], C. Sharmeela[2], and P. Sanjeevikumar[3]

[1] Vestas Technology R&D Chennai Pvt Ltd, India
[2] Anna University, India
[3] Aarhus University, Denmark
E-mail: sivaramanp@ieee.org; sharmeela20@yahoo.com;
sanjeevi_12@yahoo.co.in

Abstract

Electricity places an important role in the modern world. Transformers are one of the key piece of equipment in modern power systems. Because of various reasons, the failure of transformers can happen unexpectedly, resulting in a power supply outage to the end-users. Distributed generation (DG) allows the bidirectional power flow in the system with higher integration of distributed renewable energy sources (RES), especially solar photovoltaics. It is possible to install the solar PV system in the range between 10% and 120% of the distribution transformer rating under the DG. Failures of distribution transformers affect the power supply to end-users and the DG integration into the distribution grid. Consistent monitoring of distribution transformers by means of physical inspection is one of the major concerns of distribution companies/end-users to prevent unexpected failure. Currently, distribution transformer parameters like state of working (i.e., online/offline) is monitored, and other parameters like oil level, oil temperatures, winding temperature, etc., are not monitored from the remote end. This chapter discusses the monitoring of health conditions of a laboratory prototype 250-VA single phase oil-immersed transformer using Internet of Things (IoT) technology to achieve the conventional distribution systems toward a smart distribution system.

Keywords: Distribution transformer, distribution system, distributed generation (DG), Internet of Things (IoT), smart grid.

4.1 Introduction

Conventionally, electricity is generated by large power plants at remote locations and the power is then transferred to the load center [3]. Power flow in the conventional power system is unidirectional, and transformers here play an important role. The modern electric power systems allow bidirectional power flows with smart communication devices. The concept of distributed generation (DG) introduced localized power generation closer to the load center and it allows bidirectional power flow into the system. It is possible to install the solar PV system in the range between 10% and 120% of the distribution transformer rating under the DG [22]. Failures of distribution transformers affect the power supply to end-users and the DG integration into the distribution grid. The end-users always need high reliability in input power supply from the distribution company for trouble-free operation of their loads. The supply's reliability depends on the performance of the equipment in the system, particularly transformers [23–25]. An unexpected failure of distribution transformers results in power outages to the end-users, reducing the reliability of the power supply. Usually, power or distribution transformers have a 20–35-year design lifetime, and this can be extended up to 60 years with good maintenance in practice. Failure of distribution transformers is common because of poor maintenance, overloading and harmonics, lightning over current/voltages, internal/external loose connections, and internal/external short circuits. Hence, monitoring transformers are essential for high supply reliability to the end-users in the distribution system towards the smart distribution system. Presently, PLC-based systems are used for monitoring transformers locally at the site location. In a distribution system in metro cities, many distribution transformers are in place/operation, and it is difficult to monitor all of them locally. Hence, it is essential to monitor all the distribution transformers in a common/centralized location. An Internet of Things (IoT) based technology enables the monitoring of many distribution transformers used in the distribution system at a commonplace, i.e., central monitoring control center or main control center. An IoT-based system has been designed so that it will continuously monitor the essential parameters of the transformers throughout its day-to-day operation. The controller is used to compare the various measured parameters (like the voltage, current, oil level, or temperature) with the reference values. If any values exceed the reference value, it will give the notification and/or alarm to prevent damage/failure/tripping.

4.2 Introduction to the Transformer

A transformer is static electrical equipment, and it is used to transfer energy between two or more circuits using electromagnetic induction. The transformers transform the voltage level from one voltage to another voltage at the same operating frequency [2, 3]. It comprises two windings, namely primary windings and secondary windings [7]. These two windings are not electrically connected but are connected through the magnetic core. It works on the principles of Faraday's law of electromagnetic induction.

The main parts of the transformer are as follows:

1. core (made up of silicon steel);
2. the primary winding (made up of copper or aluminum);
3. secondary winding (made up of copper or aluminum.

The installation of a distribution transformer of 100 kVA, 11 kV/0.433 kV, 50 Hz used to power the residential and agriculture loads in a rural village in India is shown in Figure 4.1.

Figure 4.1   The installation of 100-kVA distribution transformer.

4.3 Failure of the Distribution Transformer

The life of the distribution transformer is significantly affected/reduced due to overloading [6]. This results in unexpected tripping and/or failures of the distribution transformer. Hence, it leads to power supply interruption to various end-users connected to it, and the system's reliability is questionable [8]. The most common causes for failure of distribution transformer are [13–15] as follows:

1. overloading;
2. improper maintenance;
3. electrical faults (both internal and external);
4. overvoltage is due to lightning surge;
5. loose connections;
6. insulation degradation;
7. environmental factors like moisture, etc.;
8. overheating because of harmonics and unbalanced loading;
9. accident to its physical infrastructure by vehicles, etc.;
10. accident to its physical infrastructure by vehicles, etc.;
11. presence of oxygen;
12. solids in the insulating oil.

The failure and causes of distribution transformer in Punjab State Power Corporation Limited in India between 2010 and 2015 are given in Figure 4.2. Three hundred and forty-eight distribution transformers failed in this period [1].

In order to avoid the unexpected failure or outage of the distribution transformer, it is essential to monitor the distribution transformer and maintain them with care.

4.4 Transformer Health Monitoring System through IoT

The advantages of monitoring the transformer through IoT are as follows [4, 5].

- Monitoring many distribution transformers manually is difficult, and IoT-based monitoring is more convenient for distribution operators or companies.
- More reliable as compared to local monitoring.
- Economically, IoT-based monitoring is cheaper as compared with local monitoring.

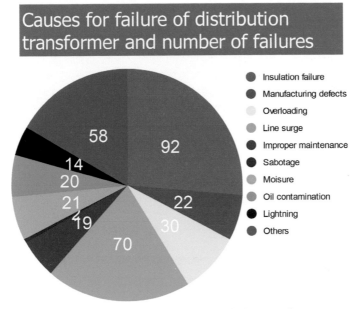

Figure 4.2 Failure and causes of distribution transformer.

Figure 4.3 shows the typical block diagram of IoT-based health monitoring of oil-immersed distribution transformer.

The IoT-based system monitors the oil-immersed transformer essential parameters throughout its operation continuously. They are oil temperature, winding temperature, oil level, voltage, and current flow. The sensors are used to measure these parameters and communicate the measured values to the microcontroller.

4.4.1 Winding and Oil Temperature Sensor

Due to the overloading of the transformer, the winding temperature and oil temperature get increased, resulting in transformer failure or outage. Hence, it is essential to monitor the temperature inside the transformer [20]. If winding or oil temperature increases above the allowed temperature, it is mandatory to reduce the temperature to avoid transformer failure or outage. The temperature sensor is used to measure the temperature inside the transformer cubicle, i.e., winding temperature and oil temperature. This sensor will measure the transformer winding and oil temperature and communicate it to the microcontroller.

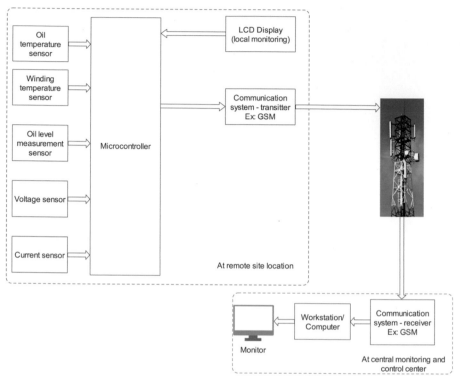

Figure 4.3 Block diagram of IoT-based health monitoring of oil-immersed distribution transformer.

4.4.2 Oil Level Monitoring Sensor

The transformer oil is generally used for two main reasons. They are insulation and cooling. Transformer oil levels will be reduced due to leakages and evaporation due to overheating. It is essential to maintain the transformer oil level at the required level during the normal operation. In case the oil level is reduced below the required level, it affects both cooling as well as insulation. So, it is not safe to operate the transformer with a lesser oil level on it. Hence, it is essential to monitor the oil inside the transformer. An ultrasonic sensor is used to measure the oil level and communicate the oil level to the microcontroller [21].

4.4.3 Current Sensor and Voltage Sensor

The current is drawn from the source through the transformer during the operation of electrical equipment (i.e., loads). Whenever there is an overloading condition, the transformer also gets overloaded. Due to overloading, the temperature inside the transformer is also increasing. Hence, it is essential to monitor the transformer loading condition. Current sensors or current transformers measure the current flowing through the transformer and give the output. Based on the current flow through the transformer, the output signal is proportional to it.

Similarly, a voltage sensor is used to measure the voltage at which the transformer operates and communicate the output signal to the microcontroller.

4.4.4 Microcontroller

A microcontroller is a device used to compare the various measured quantities, such as oil temperature, winding temperature, oil level, voltage, and current with the pre-defined or reference values. It will communicate with a local LCD for local monitoring and remote monitoring through IoT. If any of the measured values exceeds the reference value, it will initiate the alarm and/or notification.

4.4.5 LCD or Monitor

The LCD or monitor is used as a human-machine interface for displaying the various measured quantities for human understanding. The LCD or monitor is used for local monitoring at the remote site location.

4.4.6 Communication System

A communication system transfers the various measured quantities, like oil level, winding temperature, etc., to a central monitoring location from a remote site location. Many proven wireless communication technologies are available in the literature to transfer the data between two or more devices [9–12]. Some of the wireless communication technologies for long-distance communications are as follows:

1. Global system for mobile communication (GSM)
2. General packet radio service (GPRS)
3. Global positioning system (GPS)

The GSM is a mobile phone communication system used to communicate between multiple devices [16–19]. It is used to communicate the remote site location measuring devices with central monitoring and control centers. GSM acts as a transmitter at remote site locations and transfers the measured quantities to central monitoring. In central monitoring and control location,

GSM acts as a receiver and will receive the signal from the remote site location.

4.4.7 Central Monitoring and Control

The central monitoring and control center are commonplace to monitor the various equipment conditions and their parameters, as shown in Figure 4.3. It received the measured quantities from various equipment in multiple remote site locations. It displays the various measured parameters of each equipment per remote site location for easy user understanding. If any measured quantities exceed the reference value, it will initiate the alarm and/or notification. The central monitoring and control center operator must acknowledge the alarm/notification and necessary action should be taken to prevent the transformer failure.

4.5 Case Study

It showed the block diagram of the laboratory setup of IoT-based transformer monitoring system in Figure 4.4.

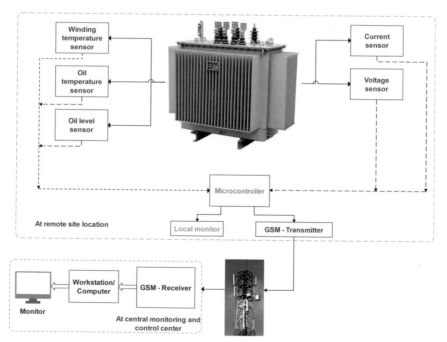

Figure 4.4 Block diagram of IoT-based transformer health monitoring.

Figure 4.5(a) shows the experimental hardware setup, and the single-phase transformer is shown in Figure 4.5(b).

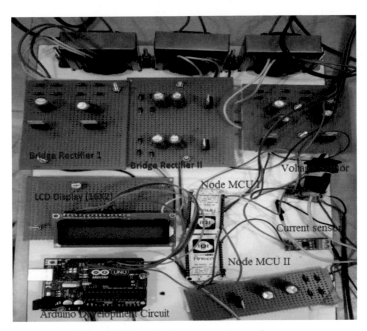

Figure 4.5(a)   Hardware experimental setup.

Figure 4.5(b)   Single phase transformer.

The monitoring of various transformer parameters through IoT for normal operation is shown in Figure 4.6(a) and that of high winding and oil temperature is shown in Figure 4.6(b).

Normal Oil Temperature	35 ℃
Winding Temperature	37 ℃
Voltage	±220 to 240 Volts
Current	Max 12 Amps

oil temperature sensor	Normal Temp
winding temperature sensor	Normal Temp
Oil Level Sensor distances	298cm
Current Sensor	-7.71 mAmps
Voltage Sensor	234.04 Volts
Date	2018-05-14
Time	2:03:25pm

Figure 4.6(a)   Online screenshot of the web page – normal operation.

Normal Oil Temperature	35 ℃
Winding Temperature	37 ℃
Voltage	±220 to 240 Volts
Current	Max 12 Amps

oil temperature sensor	High Temp
winding temperature sensor	High Temp
Oil Level Sensor distances	98cm
Current Sensor	10.14 mAmps
Voltage Sensor	231.55 Volts
Date	2018-05-14
Time	3:15:29pm

Figure 4.6(b)   Online screenshot of the web page – high winding temperature.

The system operator must acknowledge the alarm/notification; necessary preventive action must be taken to avoid the transformer failure.

4.6 Conclusion

The IoT-based health monitoring of the transformer is presented in this chapter. The essential parameters such as oil level, winding temperature, oil temperature, ambient temperature rise, and transformer loading can be monitored continuously throughout its operation. If any parameter exceeds the reference value, it will give the abnormality alarm/notification. Necessary preventive action needs to be taken immediately to prevent the transformer from failing. The IoT-based method is very convenient for monitoring many distribution transformers in urban distribution systems in metro cities like Chennai. As compared with manual monitoring, this method is much useful and reliable.

References

[1] J. Singh, S. Singh, and A. Singh, Distribution transformer failure modes, effects, and criticality analysis (FMECA), Engineering Failure Analysis, pp. 180–191, vol 99, 2019.

[2] R. Murugan and R. Ramasamy, Failure analysis of power transformer for effective maintenance planning in electric utilities, Engineering Failure Analysis, pp. 182–192, vol 55, 2015.

[3] P. Sivaraman, C. Sharmeela, R. Mahendran, and A. T. Nayagi, Basic Electrical and Instrumentation Engineering, John Wiley & Sons, 2020.

[4] A.-R. AI-Ali, A. Khaliq, and M. Arshad, GSM-based distribution transformer monitoring system, IEEE MELECON 2004, Croatia, May 12–15, 2004, pp. 999–1002, vol. 3.

[5] D. Duttachowdhury, V. Patil, A. Parab, R. Patel, and K. Niikum, Transformer monitoring and control using IoT, IOSR Journal of Engineering, pp. 40–43, vol. 10, Mar. 2018.

[6] R. Singh and A. Singh, Causes of failure of distribution transformers in India, 9th International Conference on Environment and Electrical Engineering (EEEIC 2010), 2010, pp. 388–391.

[7] P. Sivaraman and C. Sharmeela, Introduction to electric distribution system. In B. Kahn, H.H. Alhelou, & Ghassan (Eds.), Handbook of Research on New Solutions and Technologies in Electrical Distribution Networks, pp. 1–31. Hershey, PA: IGI Global, 2020.

[8] P. Sivaraman and C. Sharmeela, Existing issues associated with electric distribution system. In B. Kahn, H. H. Alhelou, & Ghassan (Eds.), Handbook of Research on New Solutions and Technologies in Electrical Distribution Networks, pp. 1–31, Hershey, PA: IGI Global, 2020.

[9] B. Li, B. Zhang J. Guo, and H. Yao, Study of cognitive radio based wirelesss access communication of power line and substation monitoring system of smart grid, International Conference on Computer Science and Service System, IEEE, 2012.

[10] P. Sivaraman and C. Sharmeela, IoT-based battery management system for hybrid electric vehicle, In A. Chitra, P. Sanjeevikumar, J.B. Holm-Nielsen, & S. Himavathi (Eds.), Artificial Intelligent Techniques for Electric and Hybrid Electric Vehicles, pp. 1–16, Beverly, MA: Scrivener Publishing LLC, 2020.

[11] L. Atzori, A. Iera, and G. Morabito, The Internet of Things: A survey, Computer Networks, pp. 2787–2805, vol 54, 2010.

[12] N. Harish, V. Prashal, and D. Sivakumar, IOT based battery management system, International Journal of Applied Engineering and Research, pp. 5711–5714, vol 13, 2018.

[13] P. Rawal, J. Pratipalsinh, and V. Devdhar, Distribution transformer failure analysis in Gujarat DISCOM, International Journal of Latest Technology in Engineering, Management & Applied Science, pp. 48–51, vol 6, Jun. 2017.

[14] R. Gayathri, S. Nanthini, M. Maheshwari, and S. Akila, Transformer health monitoring system based Internet of Things, International Journal of Pure and Applied Mathematics, pp. 959–964, vol 119, no. 15, 2018.

[15] S. Karthik, B. Ezhilkumaran, V. Vaibhav, and S. Kishorekumar, Real time health monitoring system of transformer using IoT, International Journal of Pure and Applied Mathematics, pp. 921–925, vol 119, no. 15, 2018.

[16] N. Sharma, M. Shamkuwar, and I. Singh, The history, present and future with IoT, Intelligent Systems Reference Library, pp. 27–51, 2018.

[17] A.A. Sonune, M.S. Talole, S.V. Sonkusale, A.A. Akotkar, P.S. Jaiswal, and V.N. Gayki, Condition monitoring of distribution transformer using IoT, International Journal of Engineering Research & Technology, pp. 335–338, vol 9, no. 06, Jun. 2020.

[18] A.K. Gautam, P. Gharpure, and P.V. Gawande, GSM based remote monitoring of parameters of transformer, International Journal of Modern Trends in Engineering and Research, pp. 409–412, vol 03, no. 03, Mar. 2016.

[19] C.D. Oancea, GSM infrastructure used for data transmission, 7th International Symposium on Advanced Topics in Electrical Engineering (ATEE), May 2011, pp. 1–4.

[20] P. Kore, V. Ambare, A. Dalne, G. Amane, S. Kapse, and S. Bhavarkar, IoT based distribution transformer monitoring and controlling system, International Journal of Advance Research and Innovative Ideas in Education, pp. 122–126, vol 5, no. 2, 2019.

[21] P. Dimpal, P. Mayank, G. Pratik, K. Pramod, and D. Kshitija, Monitoring and controlling of transformer using IoT, International Journal of Advance Research and Innovative Ideas in Education, pp. 1250–1256, vol 6, no. 5, 2020.

[22] P. Sivaraman, C. Sharmeela, and D.P. Kothari, Enhancing the voltage profile in distribution system with 40 GW of solar PV rooftop in Indian grid by 2022: A review, 1st International Conference on Large Scale Grid Integration of Renewable Energy in India, Sep. 2017.

[23] P. Sivaraman and C. Sharmeela, Power quality and its characteristics, In P. Sanjeevikumar, C. Sharmeela, & J. B. Holm-Nielsen (Eds.), Power Quality in Modern Power Systems, Cambridge, MA: Academic Press, 2021.

[24] P. Sivaraman and C. Sharmeela, Power system harmonics, In P. Sanjeevikumar, C. Sharmeela, & J. B. Holm-Nielsen (Eds.), Power Quality in Modern Power Systems, Cambridge, MA: Academic Press, 2021.

5

Introduction To Machine Learning Techniques

Saniya M. Ansari[1], Ravindra R. Patil[2], Rajnish Kaur Calay[3], and Mohamad Y. Mustafa[3]

[1]E & TC Department, Dr D Y Patil School of Engineering (DYPSOE), India
[2]PhD Scholar, Department of Building, Energy and Material Technology, UiT The Arctic University of Norway, Norway
[3]Department of Building, Energy and Material Technology, UiT The Arctic University of Norway, Norway
E-mail: saniya.ansari@dypic.in; ravindra.r.patil@uit.no; rajnish.k.calay@uit.no; mohamad.y.mustafa@uit.no

5.1 Why and What is Machine Learning?

Machine learning (ML), an integral part of artificial intelligence (AI), has emerged as one of the most important technologies in today's world. So what do these terms mean? In this chapter, the concept and involved techniques in ML have been elucidated.

AI has the potential to make decisions like humans and consists of standard rules encoded in the form of computer programs. Computers and computational techniques are included in the central field of computer science. The ML algorithms learn from the processed data and build predictions on that data. It is possible to alter the action and reaction of large data in ML to achieve higher adaptability, efficiency, and scalability. Next, deep learning is also a specific kind of advanced ML which uses deep neural networks to solve distinct problems. The deep neural network implements minuscule computations on numerous layers for performing duties like humans [1]. The American pioneer in the AI and computer gaming area, Arthur Samuel, first introduced the ML phrase in 1959 and defined it as follows: "it gives computers the ability to learn without being explicitly programmed."

Then, Tom Mitchell in 1997 defined ML as, "A computer program is said to learn from experience E concerning some task T and some performance measure P, if its performance on T, as measured by P, improves with experience E."

5.1.1 Phrases in Machine Learning

- **Model** – It is the key element of ML technology and is trained by utilizing an ML algorithm to generate outturns.
- **Algorithm** – It is a bunch of rules and computational techniques to return meaningful details.
- **Training data** – It is the crucial aspect in the ML algorithms which comprises features, patterns, and key trends.
- **Testing data (TD)** – It helps to assess the accuracy of the trained model.
- **Predictor variable (PV)** – It is an attribute of the data which can forecast the outturn.
- **Response variable (RV)** – It is an attribute of the output variable and the PV(s) should predict it.

5.1.2 Steps Involved in Machine Learning Practices

The ML practices consist of the following steps:

- defining the problem statement and following the objectives;
- collecting the data;
- processing the data to return meaningful information;
- selecting algorithms;
- training of the model;
- assessment and optimization of the model;
- tuning of the algorithm;
- generating outturns.

5.1.3 Properties of Data

The detailing of data properties is listed in Table 5.1.

Table 5.1   Detailing of data properties.

Data Property	Detailing
Velocity	Data creation and streaming rate
Volume	Data scale
Veracity	Assurance and truthiness in data
Variety	Contrasting types of data images, audios, text, videos
Value	Significance of data in the form of details

5.1.4 Real-World Applications of Machine Learning

Figure 5.1 presents the real-world applications of machine learning techniques.

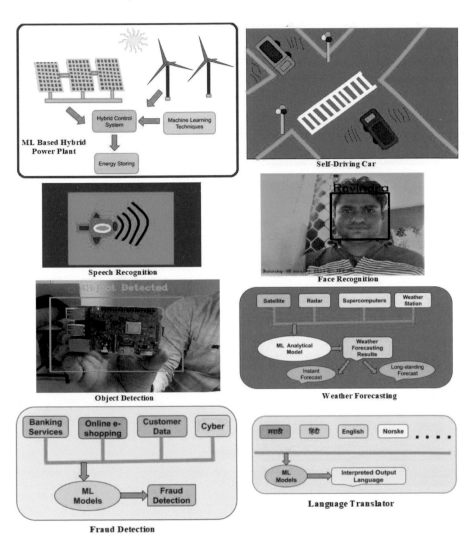

Figure 5.1 Real-world applications of machine learning.

5.2 Classification of Machine Learning Techniques

ML techniques are classified into three significant types to be contingent on the mode of the learning [5]. The ML techniques are as follows:

- supervised learning;
- unsupervised learning;
- reinforcement learning (RL).

Let us dive deep into the classified techniques of ML as shown in Figure 5.2.

5.2.1 Supervised Learning

It is the classified learning technique of ML which learns from the proficiently "Labeled" data to yield the output. It means that input data have beforehand labeled with intended output. It comes to know that labeled data given to the algorithms play a crucial role in training and contain all features to generate the final output [8]. This data is nothing to a supervisor in the learning process and similar to the student who acquires knowledge from the teacher's supervision.

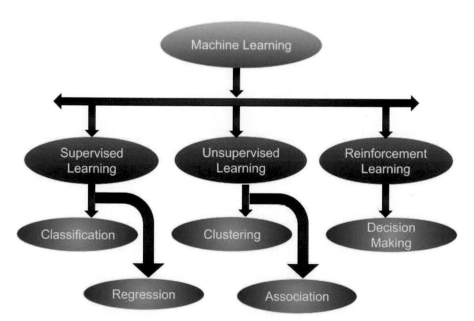

Figure 5.2   Classification of machine learning techniques.

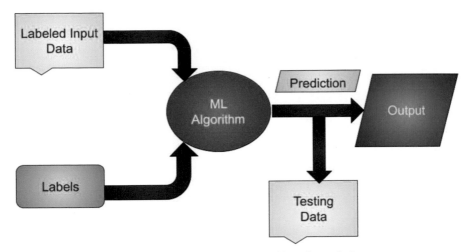

Figure 5.3 Workflow of supervised learning technique.

The supervised learning algorithms have a goal of coming upon a mapping function to map the input variable (x) and the output variable (Y) [3]. The following equation shows the correlation:

$$Y = f(x). \tag{5.1}$$

In the realistic scenario, supervised learning is applied for different applications such as classification of images, fraud detection, assessment of risk, prediction of the score, spam email detection, etc. The workflow of the supervised learning is given in Figure 5.3.

In an example, consider we have an image of distinct types of flowers and the supervised learning model has a goal to recognize the flowers and classify them appropriately. For this, we will train the ML model for different features of flowers and will test on the new dataset (testing dataset) of flowers. So, ML trained model by a suitable algorithm will identify the flower and predict the outturn.

There are two types of supervised ML algorithms.

5.2.1.1 Classification

If an output variable is categorical, then a classification algorithm is applied to get intended results. This means it consists of classes for respective data [9, 10].

In classification, take a look at the following equation:

$$y = f(x). \tag{5.2}$$

Here, y is the categorical output.

This means y (discrete output function) is mapped to x (input variable). Following are some crucial classification algorithms:

- logistic regression;
- random forest;
- support vector machines;
- decision trees.

5.2.1.2 Regression

For relating input variable and output variable, the regression algorithms come into services and are applied for the prediction of continuous output variables (output is always real value) [11]. Following are some favored regression algorithms:

- linear regression;
- non-linear regression;
- Bayesian linear regression;
- regression trees.

The difference between the classification algorithm and the regression algorithm is given in Table 5.2.

Table 5.2 Difference between classification algorithm and regression algorithm.

Classification algorithm	*Regression algorithm*
• These algorithms are utilized with discrete data.	• These algorithms are utilized with continuous data.
• It records the input value (a) per the discrete output variable (b).	• It records the input value (a) per the continuous output variable (b).
• In this, it is tried to find out the decision boundary to classify the dataset.	• In this, it is tried to find out the best fit line to predict the right output.
• It is divided into binary class classifiers.	• It is divided into linear and non-linear regression.
• In this, the output variable must seem to be a discrete value.	• In this, the output variable must seem to be real value.

5.2.2 Unsupervised Learning

In an unsupervised learning technique, algorithms learn from simple examples and do not depend on certain related responses. An algorithm itself determines the insight and hidden patterns from data. It is the same as a human control system to learn newer things. In this way, it can be said that an unlabeled dataset is used to train the models and permitted to proceed with no supervision on data [12]. The regression and classification problem cannot be solved plainly and it is only due to the unavailability of output data.

Overall, the unsupervised learning technique has an objective to search out the basic pattern of the dataset, group the data based on equality, and record the respective dataset in compressed format. The workflow of unsupervised learning techniques is given in Figure 5.4.

In an example, consider we have an image of different types of flowers and the objective is to identify flowers appropriately. For this, an unsupervised learning algorithm will solve this problem by clustering the imagery dataset of flowers into groups as per image similarities.

Unsupervised learning is used due to the following points.

- It helps to find important insights and hidden patterns from data.
- It is similar to the critical thinking of humans to learn newer things from experiences.
- It works on non-labeled and non-categorized data and it is the function why it is called an unsupervised learning technique.
- In a realistic scenario, it is not possible to have input data always with correlating output. So, unsupervised learning helps to solve such tasks.

There are also two types of unsupervised ML algorithms.

5.2.2.1 Clustering

In clustering, the objects are grouped into clusters such as one group of objects with higher feature similarities and one new group of objects with lower or no feature similarities.

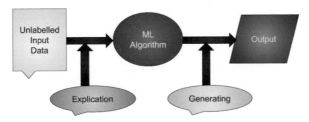

Figure 5.4 Workflow of unsupervised learning technique.

Simply, clustering analyzes the feature similarities between the objects and makes a category of inherent data depending on it. There are different types of clustering based on soft and hard clustering methods such as density-based, fuzzy, distribution model-based, hierarchical, partitioning type, etc. [13].

5.2.2.2 Assoclation

Association is used to find correlation within dataset variables. This means the bunch of things that seem jointly in the dataset is determined by association [14]. It is nothing but finding a rule which reports about the dataset.

The analysis of the market basket (a person wants to purchase A as well as B) is the best example of association.

Following are some favored unsupervised learning algorithms.

- KNN (*k*-nearest neighbors);
- anomaly detection;
- *K*-means clustering;
- principle component analysis;
- neural networks.

Table 5.3 presents the advantages and disadvantages of both supervised and unsupervised learning, as well as the difference between them, which is given in Table 5.4.

5.2.3 Reinforcement Learning

This ML technique depends on feedback and its smart factor or agent (computer program) learns from prior experiences obtained by performing movement to deal with the surroundings. For good movement, the factor obtains positive feedback, whereas, for worst movement, it obtains negative feedback [2].

Overall, it learns from trial-and-error experiences acquired from realistic scenarios to perform excellently. It can be said that the smart factor (computer program) relates to the surroundings and learns from it to move.

The AI smart factor or agent is concerned with RL where it learns automatically from earlier experiences without any human perception [15].

The AlphaGo and self-driving cars are the best state-of-the-art ML examples of RL and comprise decision-making sense.

The smart factor or agent consists of three tasks such as acting, changing, or remaining in the same state and obtaining feedback. These tasks are used to inspect and learn about the surroundings. For positive feedback and negative

Table 5.3   Advantages and disadvantages of supervised learning and unsupervised learning.

Supervised learning	*Advantages*	• It helps the model to predict the outruns based on earlier experiences. • It solves mostly occurring real-world issues. • In this, we get a precise idea regarding object classes.
	Disadvantages	• It cannot handle critical tasks. • The computation time is large for model training. • It is not able to predict right outruns if the test and train data are different.
Unsupervised learning	*Advantages*	• It handles critical tasks because it does not comprise input labeled data. • It is favored mostly because unlabeled data is available simply as compared to labeled data.
	Disadvantages	• It is a hard technique as compared to supervised learning as it does not comprise consistent outturns. • The results might not seem so much correct due to no pre-idea about outturns.

Table 5.4   Difference between supervised learning and unsupervised learning.

Supervised learning	*Unsupervised learning*
• It is computationally very complex.	• It is computationally less complex.
• The input data is known and labeled.	• The input data is unknown and unlabeled.
• It is classified into two types of problems as classification and regression.	• The clustering and associations are two types of problems in unsupervised learning.
• It requires supervision to learn the model.	• It does not require supervision to learn the model.

feedback, the smart factor or agent obtains a positive point and negative point, respectively.

5.2.3.1 Crucial terms in reinforcement learning

a) **Smart Factor or Agent () –**
   The unit discerns or inspects the surrounding and make movement on them.

b) **Environment () –**
   A place inside which smart factor or agent is available or near.

c) State () –

After the movement of smart factor or agent, the environment gives back its place.

d) Action () –

The movements of the smart factor or agent inside the environment.

e) Policy () –

It is the plan of the action appealed by the smart factor or agent for the following action or movement on the present state.

f) Reward () –

It is the feedback given to the smart factor or agent by the environment or surrounding to assess the movement of smart factor or agent.

g) *Q*-value () –

It is similar to the value; even so, it catches one extra parameter like current action or movement (a).

h) Value () –

It is inverse to the short-term reward and looked for long-term return.

5.2.3.2 Salient features of reinforcement learning

- It is a trial and error based procedure.
- The smart factor or agent may obtain a setback reward.
- There is no need to teach the smart factor or agent about surroundings or the environment as well as what movement is required to proceed.
- In this, the smart factor or agent requires to automatically inspect the surrounding or the environment to obtain the greatest positive rewards.
- The value-based, policy-based, and model-based are three methods to execute RL.

5.2.3.3 Types of reinforcement learning

Primarily, there are two classes of RL as given below.

a) Positive Reinforcement

Positive reinforcement means to join some kind to enhance a propensity so that an expected behavior will be taking place anew. It affects the smart factor's or agent's behavior conclusively as well as enhances its robustness.

This positive reinforcement can continue the alterations for large spam; even so, maximum positive reinforcement may require a burden of places which can minimize the outturns.

Table 5.5 Difference between supervised learning and reinforcement learning.

Supervised learning	Reinforcement learning
• It functions on the occurring dataset.	• It functions by interrelating with the surrounding or environment.
• There is the availability of a labeled dataset.	• No labeled dataset is available in it.
• When the input is present, then decisions are produced.	• In this, decisions are reserved sequentially.
• In this, the algorithm is trained to predict the outturns.	• There is no earlier training given to the smart factor or agent.

b) Negative Reinforcement

It is inverse to the positive reinforcement and newly enhances the tendency of coming behavior by keeping negative terms away.

It is highly productive than positive reinforcement and based on behavior as well as places.

The difference between supervised learning and RL is given in Table 5.5.

5.2.3.4 Reinforcement learning algorithms

There are some key RL algorithms as given below [16].

a) Q-Learning (Q Stands for Quality)

It is an off-policy RL algorithm and is utilized for the learning of temporal differences (TDs).

In this, $Q(s, a)$ is a value function that specifies in what way the good action "a" should be taken at a specific "s" state.

b) Deep Q Neural Network (DQN)

It is Q-learning by comprising neural networks. It is utilized for the surrounding of big state space.

c) State Action Reward State Action

It is an on-policy RL algorithm and is utilized for the learning of TDs.

The state action reward state action (SARSA) is applied to compute $Q\pi(s, a)$ intended for recently proposed policy π and all (s-a) pairs.

The name SARSA is given due to using of $Q(s, a, r, s', a')$.

Here, s = initial state, a = initial action, s' = new state, a' = new action, and r = reward.

d) Temporal Difference

It is a method to learn about the prediction of an extent that rests on imminent signal values.

To forecast an amount of the entire quantity of reward predictable over the imminent signal, the TD algorithm is utilized.

5.3 Some Crucial Algorithmic Mathematical Models in Machine Learning

The crucial mathematical models of ML are specified below [4].

5.3.1 Logistic Regression

It is consumed to forecast the variable reliant on category by utilizing a specified set of self-regulating variables. The values of categorical variables are predicted by a logistic regression algorithm [17]. The *S*-curve is used to classify the distinct samples. The logistic regression has three types based on categories such as binomial, multinomial, and ordinal.

The outturn of logistic regression lies between 0 and 1 as shown in Figure 5.5, and this algorithm is applied in the state of probability requirement within two classes.

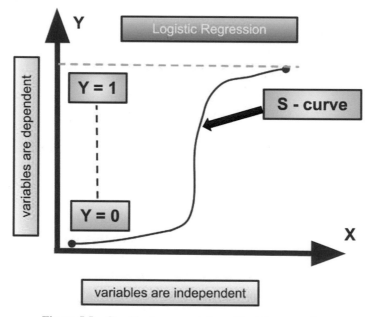

Figure 5.5 Graphical representation of logistic regression.

- **Sigmoid Function:**

This mathematical function is applied for mapping the predicted values to probabilities. In this, assessment is done for real value into a different value on a scale of 0 and 1. So, it is said that this algorithm is based on the threshold value. The logistic regression values do not go across the boundary, and, due to this, the *S*-form of the curve is structured and known as the logistic function or sigmoid function.

- **Theories:**

The dependent variable must remain a categorical kind. The multi-collinearity should not be applicable for the independent variable.

- **Mathematical Equation:**

This equation is gained from the linear regression equation. We have the straight-line equation as follows:

$$y = b0 + b1 \times 1 + b2 \times 2 + \cdots + bn \times n. \tag{5.3}$$

Now divide the above equation by 1 - *y*, and we get the following:

$$y / (1 - y) \,;; \ 0 \text{ for } y = 0 \text{ and infinity for } y = 1. \tag{5.4}$$

Taking logarithm on the above equation, we get the resulting equation as follows:

$$\log [y / (1 - y)] = b0 + b1 \times 1 + b2 \times 2 + \cdots + bn \times n. \tag{5.5}$$

5.3.2 Decision Trees

It is utilized for both regression and classification problems but is typically favored for classification problems. It contains two nodes as the decision node and leaf node.

These inner nodes signify the dataset's features in which the decision rules are denoted by branches and outturns by leaf node [18]. Figure 5.6 shows the interconnection inside the decision tree technique.

This workflow of the algorithm is as given below.

- Start tree through *S* as a root node covering the comprehensive dataset.
- By making use of the attribute selection measure (ASM), search for the finest dataset attribute.
- *S* is divided into subsets comprising probable values intended for the finest attributes.

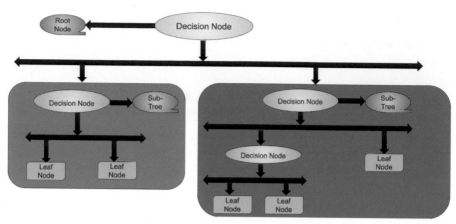

Figure 5.6   Decision tree diagram.

- Make a node of the decision tree for covering the greatest attribute.
- Create newer decision trees by utilizing an earlier initiated subset of a given dataset and repeat this process till getting the final node which cannot be classified further which is nothing but leaf node.

How to select attribute for root and sub-nodes?

The attribute selection technique (ASM) is utilized for a selection of the finest attribute for root and sub-nodes. It has two types as given below with mathematical equations.

- **Information Gain:**

It computes the total of information about the class provided by the feature.
The formula for calculating information gain is as follows:

$$\text{Information Gain} = \text{Entropy} -$$
$$[(\text{Weighted Average}) * \text{Each Features Entropy}]. \qquad (5.6)$$

Here, entropy is given by

$$\text{Entropy} = [(-P\,(\text{Yes})\,\log 2\,P\,(\text{Yes})) - (P\,(\text{No})\,\log 2\,P\,(\text{No}))]. \quad (5.7)$$

Entropy is denoted by S - total number of samples; $P\,(\text{Yes})$ – the probability of Yes; $P\,(\text{No})$ – the probability of No.

- **Gini Index:**

It measures purity or impurity utilized through generating a decision tree in the algorithm of classification and regression tree (CART).

It is computed by the below formula as

$$\text{Gini Index} = 1 - \sum j \ Pj \ 2. \qquad (5.8)$$

This algorithm is easy and valuable for decision-related issues and involves less data cleaning.

5.3.3 Linear Regression

It is consumed to forecast the variable reliant on a continuous state by utilizing a specified set of self-regulating variables. The values of continuous variables are predicted by a linear regression algorithm [11].

Simply, it clarifies the connection within variable which generally seems in the format of one dependent and one or more independent. The graphical representation of linear regression is given in Figure 5.7.

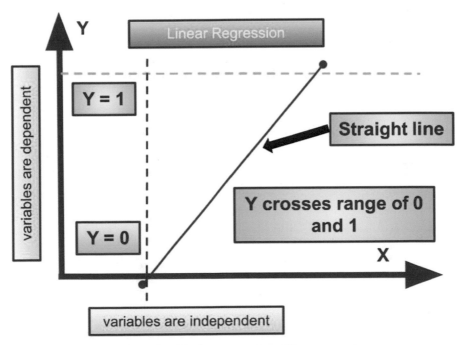

Figure 5.7 Graphical representation of linear regression.

- **Mathematical Equation:**

The basic equation of regression is given as follows:

$$y = c + b * x. \tag{5.9}$$

Here, b = coefficient of regression, c = constant, x = score of proceeding independent variable, and y = score of projected dependent variable.

It is of two types, namely simple linear regression and multiple linear regression.

The linear regression line represents the correlation between the dependent and independent variables. Figure 5.8 signifies about positive linear relationship, whereas Figure 5.9 depicts a negative linear relationship.

5.3.4 K-Nearest Neighbors

It is a non-parametric algorithm and it reflects that there are no assumptions completed on primary data. The lazy learner algorithm is also another name of the KNN algorithm.

In the training stage, the KNN algorithm impartially gathers the dataset, and after receiving newer data, it is classified into the same class alike to the received data. It works well on noisy as well as large data [19, 20].

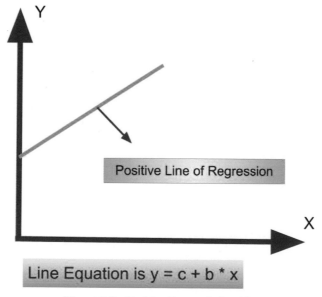

Figure 5.8   Positive linear relationship.

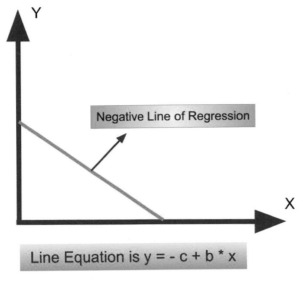

Figure 5.9　Negative linear relationship.

This workflow of this algorithm is as given below.

- Calculate the Euclidean distance as of the question instance toward the labeled instance.
- The labeled instances are ordered by rising distance.
- Compute k of adjacent neighbors relying on root mean squared error.
- Compute a converse distance weighted average through the multivariate neighbors of k-nearest.

The feature space with new arising data point is given in Figure 5.10 and feature space with classified newer arising data point in Figure 5.11.

For this, Euclidian distance is calculated by the following formula:

$$\text{Euclidian Distance} = \sqrt{(x_2 - x_1)^2 + (y_2 - y_1)^2}. \qquad (5.10)$$

This equation provides the nearest neighbors for the newer data point in the feature space as described for class C and class D in Figures 5.10 and 5.11.

The selection of value K in the model is completely trial and error based. So, we have to go with some values for K, and if we get the intended results, then we fix it further. Generally, $K = 5$ is favored in distinct issues. For lower values of K, we find mostly outliers mean nosiness.

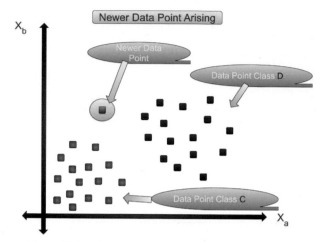

Figure 5.10   Feature space with new arising data point.

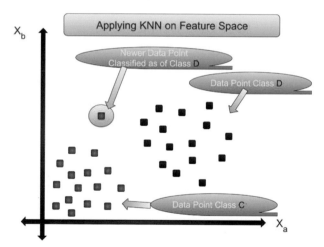

Figure 5.11   Feature space with classification of newer arising data point.

5.3.5 K-Means Clustering

In this repetitive algorithm, the unlabeled dataset is divided into k distinct types of clusters in the format that every dataset fits in the individual group which consists of the same possessions [20, 21].

It consists of the following tasks.

- Finds the finest value for K centroids by comprising iterative process.

- Allocates individual data points to neighboring k centroid. The data points closer to the typical k centroid form cluster.

Figure 5.12 shows feature space with all data points and clusters of data points are given in Figure 5.13.

Overall, the functioning of *K*-means lies in the most effective cluster formed by it. The elbow method is used to find the value of *K* or the number of

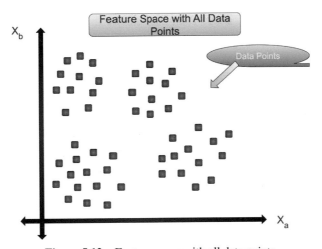

Figure 5.12   Feature space with all data points.

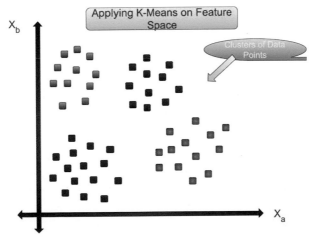

Figure 5.13   Feature space with clusters of all data points.

clusters. It utilizes the perception of within cluster sum of squares (WCSS) to describe the entire differences inside the cluster. The mathematical equation to calculate the WCSS value for four clusters is given below:

$$\text{WithinClusterSumofSquares(WCSS)} =$$
$$\sum Pi \text{ in } Cluster1 \text{ } Distance(Pi \text{ } C1)2$$
$$+ \sum Pi \text{ in } Cluster2 \text{ } Distance(Pi \text{ } C2)2$$
$$+ \sum Pi \text{ in } Cluster3 \text{ } Distance(Pi \text{ } C3)2$$
$$+ \sum Pi \text{ in } Cluster4 \text{ } Distance(Pi \text{ } C4)2. \qquad (5.11)$$

Here, $\sum Pi$ in $Cluster1$ $Distance(Pi$ $C1)2$ is the sum of the square of the distances concerning individual data point and related centroid inside the cluster and is similar for every cluster. The distance between data points can be calculated either by Manhattan distance or by Euclidean distance. That is all about K-means for solving real-world issues.

5.4 Pre-Eminent Python Libraries Intended for Machine Learning

For real-world implementation, python is a very simple and robust programming language to hack embedded platforms. There are multiple libraries available in python for ML and some of them are given below.

- SciPy
- NumPy
- Scikit-learn
- Pillow
- Pandas
- Matplotlib
- PyTorch
- Keras
- TensorFlow

The OpenCV is a very popular computer vision and ML library for the development of real-world applications. It also contains an inference module for running deep neural networks with CUDA support. It is compatible with several python libraries for ML applications on an embedded platform. So, embedded vision has become a vast research area for real-world visual applications [6]. Object detection and object tracking are now utilized in a score of fields such as autonomous navigation, military applications, security purposes, etc. [7].

Let us take some real-world examples with python programming.

5.4.1 Human Detection (OpenCV, HoG, SVM with Multi-Threading)

Human detection is one of the crucial tasks related to security and has been implemented with HoG and SVM [22].

Python code:

```
import cv2
import numpy as np
from imutils.object_detection import non_max_suppression
import imutils
from imutils.video import VideoStream
import datetime

image = cv2.imread("imagePath")
hog = cv2.HOGDescriptor()
hog.setSVMDetector(cv2.HOGDescriptor_getDefaultPeopleDetector())

while True:
    imageDetect = imutils.resize(image, width=400)
    timingdetails = datetime.datetime.now()
        timingdetails = timingdetails.strftime("%A %d %B %Y
                %I:%M:%S%p")
    cv2.putText(imageDetect, timingdetails, (10, imageDetect.shape[0] - 10),
        cv2.FONT_HERSHEY_SIMPLEX,
            0.35, (255, 0, 255), 1)

    (rectangle, weights) = hog.detectMultiScale(imageDetect,
        winStride=(16,16),
            padding=(16,16), scale=1.05)
    rectangle = np.array([[x, y, x + w, y + h] for (x, y, w, h) in rectangle])
    nms = non_max_suppression(rectangle, probs=None,
        overlapThresh=0.65)

    for (xa, ya, xb, yb) in nms :
```

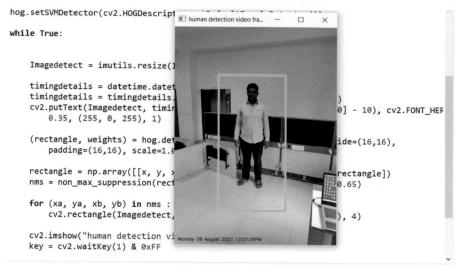

Figure 5.14   Human detection output.

cv2.rectangle(imageDetect, (xa, ya), (xb, yb), (255, 255, 0), 4)
cv2.imshow("human detection video frame", imageDetect)

key = cv2.waitKey(1) & 0xFF
if key == ord("g"):
    break
cv2.destroyAllWindows()

Output Image:
Figure 5.14 shows the output of the human detection task using OpenCV, HoG, and SVM with multi-threading for real-time consequences.

5.4.2 Instagram Filters – (OpenCV, Matplotlib, NumPy)

The interesting application of Instagram filtering is implemented below [23].

Python code:
import cv2
import numpy as np
import matplotlib.pyplot as plt
import imutils
matplotlib.rcParams['figure.figsize'] = (10.0, 10.0)
matplotlib.rcParams['image.cmap'] = 'gray'

```
threshold = 160
maximunValue = 255
framePath = "trumpGg.jpg"

frame = cv2.imread(framePath)
frame = imutils.resize(frame, 400)

src = cv2.cvtColor(frame, cv2.COLOR_BGR2GRAY)
dst = cv2.edgePreservingFilter(frame, flags=1, sigma_s=60, sigma_r=0.4)
framea = cv2.detailEnhance(dst, sigma_s=10, sigma_r=0.15)

kernelSize = 11
frame1 = cv2.GaussianBlur(src ,(3, 3),0,0)
laplacian = 10 * (cv2.Laplacian(frame1,cv2.CV_64F, ksize = kernelSize))
threshold, laplacian = cv2.threshold(laplacian , threshold, maximunValue,
cv2.THRESH_BINARY_INV)

gray, color = cv2.pencilSketch(frame, sigma_s=60, sigma_r=0.07,
shade_factor=0.05)
frameb = cv2.stylization(frame, sigma_s=60, sigma_r=0.07)

cv2.imshow("Laplacian", dst)
cv2.imshow("framea", framea)
cv2.imshow("frameb", color)
cv2.imshow("framec", gray)
cv2.imshow("framed", frameb)
cv2.waitKey(0)
cv2.destroyAllWindows()
```

Output Image:
Figure 5.15 shows the output images of Instagram filtering.

5.5 Machine Learning Techniques in State of Affairs of Power Systems

The power system is a term representing a network consisting of electric components installed to provide, move, and utilize electric power. An electrical grid is a well-known model which supplies power to industries, homes, and imperative places inside a large region [24]. The conventional grid utilizes a finite single-route interaction means power is transferred from the power

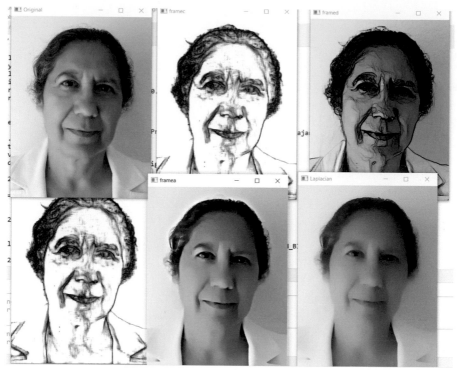

Figure 5.15 Instagram filtering images.

station to the intended area. Whereas, the smart grid operates on double-route interaction, and, in this, information and electricity are interchanged. The energy Internet is also a booming concept which involves the trend of the Internet of Things with smart grid power systems. The smart grid empowers solar and wind types of renewable energies [25], [26]. Overall, ML techniques help the modern hybrid power systems to predict and make decisions on acquired data as shown in Figure 5.16. This makes the power system most effective, credible, safe, and greener [29].

Let us have a look at some instances.

The composite inputs of time series are the major issue due to the undetermined nature of supplies of renewable energy and it invites to apply ML techniques such as convolutional neural network (CNN), multilayer perception (MLP), long short-term memory (LSTM), recurrent neural network (RNN), etc. These help to make predictions on short-term time-series in resources of renewable energy and power system [28]. Specifically, solar energy is the central resource of pure energy and is known for large electrical

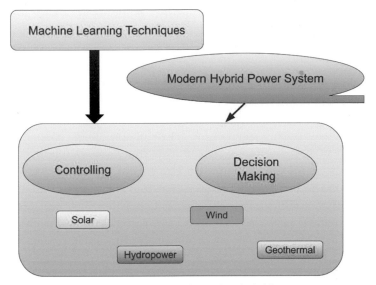

Figure 5.16 Machine learning in modern hybrid power system.

power stations related to the smart grid. The administration and steady functioning of the power system are affected by intermittency and randomness of solar energy. For this, the CNN and LSTM types of ML techniques are utilized for forecasting the photovoltaic (PV) output power correctly. These techniques have been applied for actual PV data in Belgium and Limberg [27].

The hybrid power plant consists of a wind generator and solar panels. In this, neural network type of advanced ML technique has been applied to direct the electricity allocation generated in hybrid power plant via controller [30]. The ML techniques help for the smart controlling and output predictions in power system applications [31]. At the length, the renewable modern hybrid power system is now the need in distinct human working fields.

5.6 Conclusion

In conclusion, the ML technique is the finest revolution for advancement in a score of human working fields. It is of three types such as supervised learning, unsupervised learning, and RL technique. Supervised learning resolves two kinds of problems such as classification and regression, whereas unsupervised learning functions for clustering and association issues. RL is utilized for decision-making and is classified into positive reinforcement and negative reinforcement types. Overall, there are distinct algorithms involved under the types of ML techniques for generating intended outturns as per the occurred issues. The decision trees, k-means clustering, logistic regression, linear regression, Q-value, SARSA, KNNs are the mostly utilized algorithms for solving real-world issues.

There are several applications of ML in a realistic scenario such as power systems, smart grids, energy internet, object detection, object classification, speech recognition, Google translator and map, autonomous intelligent systems, etc. Advanced ML has become a robust technique for modern hybrid power systems with renewable energy involving solar and wind energy, hydropower energy, geothermal energy, etc. Smart controlling and decision-making are the central tasks performed by ML techniques in power system applications.

References

[1] A. Voulodimos, N. Doulamis, A. Doulamis, and E. Protopapadakis. Deep learning for computer vision: A brief review. Computational Intelligence and Neuroscience, 2018.

[2] E. Alpaydin. Introduction to Machine Learning, 2020, MIT Press, Cambridge, MA, USA.

[3] T.D. Buskirk, A. Kirchner, A. Eck, and C.S. Signorino. An introduction to machine learning methods for survey researchers. Survey Practice, 11(1), 1–10, 2018.

[4] M. Kubat. An Introduction to Machine Learning, 2017, Springer, Cham.

[5] J. Brownlee, Machine Learning Mastery with Python, 527, 100–120, 2016.

[6] O.S. Vaidya, R. Patil, G.M. Phade, and S.T. Gandhe. Embedded vision based cost effective tele-operating smart robot. International Journal of Innovative Technology and Exploring Engineering (IJITEE), 8(7), 1544–1550, 2019.

[7] R.R. Patil, O.S. Vaidya, G.M. Phade, and S.T. Gandhe. Qualified scrutiny for real-time object tracking framework. International Journal on Emerging Technologies, 11(3), 313–319, 2020.

[8] S.B. Kotsiantis, I. Zaharakis, and P. Pintelas. Supervised machine learning: A review of classification techniques. Emerging Artificial Intelligence Applications in Computer Engineering, 160(1), 3–24, 2007.

[9] A. Singh, N. Thakur, and A. Sharma. A review of supervised machine learning algorithms. In 2016 3rd International Conference on Computing for Sustainable Global Development (INDIA Com), 1310–1315, 2016, IEEE.

[10] P.C. Sen, M. Hajra, and M. Ghosh. Supervised classification algorithms in machine learning: A survey and review. In emerging technology in modelling and graphics, 99–111, 2020, Springer, Singapore.

[11] D. Maulud and A.M. Abdulazeez. A review on linear regression comprehensive in machine learning. Journal of Applied Science and Technology Trends, 1(4), 140–147, 2020.

[12] M.E. Celebi and K. Aydin. Unsupervised Learning Algorithms, 2016, Springer International Publishing, Berlin.

[13] S. Kotsiantis and P. Pintelas. Recent advances in clustering: A brief survey. WSEAS Transactions on Information Science and Applications, 1(1), 73–81, 2004.

[14] A.C. Nanayakkara, B.T.G.S. Kumara, and R.M.K.T. Rathnayaka. A survey of finding trends in data mining techniques for social media analysis. Sri Lanka Journal of Social Sciences and Humanities, 1(2), 2021.

[15] S. Padakandla. A survey of reinforcement learning algorithms for dynamically varying environments. ACM Computing Surveys (CSUR), 54(6), 1–25, 2021.

[16] M.M. Afsar, T. Crump, and B. Far. Reinforcement learning based recommender systems: A survey. arXiv preprint arXiv:2101.06286, 2021.

[17] C.Y.J. Peng, K.L. Lee, and G.M. Ingersoll. An introduction to logistic regression analysis and reporting. The Journal of Educational Research, 96(1), 3–14, 2002.

[18] B. Charbuty and A. Abdulazeez. Classification based on decision tree algorithm for machine learning. Journal of Applied Science and Technology Trends, 2(01), 20–28, 2021.

[19] M. Steinbach and P.N. Tan. kNN: k-nearest neighbors. In The Top Ten Algorithms in Data Mining, 165–176, 2009, Chapman and Hall/CRC, Boca Raton, FL, USA.

[20] P. Aiwale and S. Ansari. Brain tumor detection using KNN & LLYOD clustering. IJEAT, 10(12), 187, January 2020, ISSN 2249–8948.

[21] A. Coates and A.Y. Ng. Learning feature representations with k-means. In Neural networks: Tricks of the trade, 561–580, 2012, Springer, Berlin, Heidelberg.

[22] N. Dalal and B. Triggs. Histograms of oriented gradients for human detection. In 2005 IEEE Computer Society Conference on Computer Vision and Pattern Recognition (CVPR'05), Vol. 1, 886–893, 2005, IEEE.

[23] G. Rajagopalan. Working with NumPy arrays. In A Python Data Analyst's Toolkit, 117–145, 2021, Apress, Berkeley, CA.

[24] Y.H. Song (Ed.). Modern Optimisation Techniques in Power Systems, Vol. 20, 2013, Springer Science & Business Media, Berlin.

[25] L. Cheng and T. Yu. A new generation of AI: A review and perspective on machine learning technologies applied to smart energy and electric power systems. International Journal of Energy Research, 43(6), 1928–1973, 2019.

[26] Y. Kabalci, E. Kabalci, S. Padmanaban, J.B. Holm-Nielsen, and F. Blaabjerg. Internet of things applications as energy internet in smart grids and smart environments. Electronics, 8(9), 972, 2019.

[27] G. Li, S. Xie, B. Wang, J. Xin, Y. Li, and S. Du. Photovoltaic power forecasting with a hybrid deep learning approach. IEEE Access, 8, 175871–175880, 2020.

[28] M.M. Rahman, M. Shakeri, S.K. Tiong, F. Khatun, N. Amin, J. Pasupuleti, and M.K. Hasan. Prospective methodologies in hybrid renewable energy systems for energy prediction using artificial neural networks. Sustainability, 13(4), 2393, 2021.

[29] C. Feng, M. Sun, M. Dabbaghjamanesh, Y. Liu, and J. Zhang. Advanced machine learning applications to modern power systems. In New Technologies for Power System Operation and Analysis, 209–257, 2021, Academic Press, Cambridge, MA.

[30] A. Gozhyj, P. Bidyuk, Y. Matsuki, V. Nechakhin, I. Kalinina, and O. Shchesiuk. Hybrid power plant control system based on machine learning methods. In Conference on Computer Science and Information Technologies, 251–262, 2020, Springer, Cham.

[31] K.S. Perera, Z. Aung, and W.L. Woon. Machine learning techniques for supporting renewable energy generation and integration: A survey. In International Workshop on Data Analytics for Renewable Energy Integration, 81–96, 2014, Springer, Cham.

6

Machine Learning Techniques for Renewable Energy Resources

K. Punitha*, S. Anbarasi, and T. Balasubramanian

P S R Engineering College, India
E-mail: kgpunitha@gmail.com
*Corresponding Author

Abstract

At present, renewable energy resources like solar and wind attracted abundant attention, and thanks to their green, clean, inexhaustible, and recycled nature and them being free from carbon emission, renewable energy resources are the foremost promising alternative to fossil fuels. Though renewable energy resources are accessible freely, their higher upfront price, environmental dependency, and lower efficiency act as a barrier to wider implementation. The demerits of renewable energy resources are volatility, intermittence, and uncertainty which affect the stability and reliability of large-scale renewable integration into the power generation. Hence, researchers are exploring possibilities to boost accessibility and efficiency with the help of technology such as machine learning. Deep learning, a promising kind of machine learning technique, can be incorporated with renewable energy, especially solar photovoltaic (PV) systems, in three major categories, such as forecasting, accessibility, and efficiency. Boosting the efficiency of a solar PV system requires maximum power point tracking (MPPT), which maximizes the extraction of available maximum power from PV modules. As the conventional MPPT algorithms have no prior knowledge of the maximum power point (MPP) at the beginning of the perturbation, these MPPTs demand a long convergence time to achieve MPP. The need for prior knowledge of MPP is necessary to start any conventional MPP algorithm, which the deep learning based long short-term memory (LSTM) network provides in this work. The goal of this book chapter is to implement deep learning in the solar PV system forecasting maximum voltage to provide reference value to its MPPT technique. The case study is also presented with a conclusion.

Keywords: Machine learning, deep learning, LSTM network, prediction, MPPT algorithm.

6.1 Introduction

Despite the abundance of green energy resources today, fossil fuels remain the world's primary resource of energy. Fuels that are considered fossil are hydrocarbons or their derivatives, together with natural resources like coal, petroleum oil, and natural gas. Fuels take a long time to form; so the well-known oil resource reserves are depleting faster than they can be replenished. Also they release greenhouse gases, which cause environmental change such as global warming, putting the environment at risk. Green energy has, therefore, gained a great lot of attention worldwide in recent years. A renewable energy source can be recycled into new energy in nature, such as solar power, wind energy, tidal energy, or geothermal energy. Renewable energy has two major advantages over fossil fuels. A few things to remember about renewable energy resources are that they are abundant, renewable, and inexhaustible. The second benefit is that renewable energy is carbon-free, green, and clean and, thus, benefits the environment. In particular, renewable energy can effectively reduce the emission of carbon monoxide (CO), sulfur dioxide (SO2), and dust, thereby reducing atmospheric pollution and greenhouse gas emissions. Aside from that, the use of renewable energy can diminish the need to exploit petroleum, and this will also help to protect the environment. It can reduce solid waste discharge, which reduces soil pollution. By using renewable energy, water resources are also protected by reducing the waste gases and waste liquids emitted during the process. Thus, renewable energy has become very popular in recent years [1].

In light of the rapid industrialization of our planet, it has become clear that excessive consumption of petroleum will accelerate the loss of fossil fuel reserves and harm the environment. Ultimately, this will lead to health concerns and global climate change. The wind and sun are currently the fastest-growing energy sources, along with petroleum and nuclear energy. Solar, wind, hydropower, biomass, waves, tides, and geothermal heat are reusable forms of energy to be recovered in nature. Renewable energy poses several important challenges, notably the supply of energy, due to its characteristics of sustainability and low environmental impact.

The sustainable energy refers to the inclusion of these sources into existing and future energy supply systems [2]. Energy security and regional energy shortages will be improved by developing renewable energy systems. However, this generation of various energy sources is unpredictable and chaotic as a result of the inherent instability of renewable energy and the unpredictable nature of renewable energy. So, it is still a challenge to handle with renewable energy statistics accurately. Energy monitoring with high precision can increase energy efficiency.

Developing, managing, and making energy policy all rely on energy forecasting. In the context of increased ways of supplying electricity from renewable energy sources [3], it is imperative to develop technologies for storing renewable energy. Several research papers have suggested that a number of machine learning models were used for renewable energy forecasts. This data-driven approach enables predictions of renewable energy use. A hybrid machine learning model was designed to boost renewable energy prediction accuracy. Renewable energy has been predicted for different time intervals, such as minutes, hours, days, and weeks, depending on the purpose of the prediction. Typically, predictions for renewable energy are assessed according to their accuracy and efficiency [4].

Solar irradiance reaches the earth at different rates based mostly on moving clouds. It is necessary to incorporate cloud information directly or indirectly into the formula to forecast the irradiance accurately. Physical models of clouds' generation, propagation, and extinction are difficult because clouds' generation, propagation, and extinction are stochastic. Cloud information can, therefore, usually be expressed using statistical methods [5–7]. A special focus is on forecasting irradiance during very short periods of time because clouds are persistent during that period.

A very short-term forecast is often used in large photovoltaic (PV) installations, unlike those with longer horizons whose results are critical to electricity grid operations. Maximum power point tracking (MPPT) algorithms may benefit from anticipating the potential shading of a particular section of a PV system [8]. A sub-minute forecast could also facilitate improved control of ultra-capacitors [9, 10]. The current state-of-the-art methods of forecasting very short-term irradiance were reviewed in Mane [11]. The use of sky cameras [12–14] or sensors [15, 16] can be used to analyze the data.

Several datasets related to solar engineering, such as satellite-derived irradiance measurements [17], output measurements from hundreds of solar power plants [18], and module-level data from solar power plants [19], meet the HACE principle proposed by [20] which exemplifies big data. Raw datasets have a lot of irrelevant data and noise embedded in them, which makes processing them a challenge. Not only is it clean, green, and naturally replenished in wide geography, but it also poses an unschedulable uncertainty, which threatens the stability and reliability of energy systems, especially with their large-scale integration. Considering the volatility, the intermittent nature, and random nature of renewable energy, the price of electricity production will undoubtedly increase due to increased system reserve capacity. However, renewable energy utilizes a large percentage of power electronics, therefore reducing the rotational inertia of the system and reducing the margin of stability. Sustainable energy forecasting helps to reduce uncertainty, which is vital to electrical system planning, management, and implementation [20]. The unpredictable, chaotic, and irregular nature of weather, accurate forecasting of renewable energy sources remains a challenge. Numerous algorithms exist for predicting renewable energy from 1 or 2 minutes up to one or two days ahead. The three most common categories are physical methods, statistical models, and artificial intelligence methods, along with their hybrid methods [21].

Forecasting correct energy over time is crucial for the growth of PV technology. Aspects such as how efficiently sunlight is converted into energy and how this relationship varies over time are important. The MPPT solar charge controller, which is also called a smart DC-DC converter, is essential for any solar power system to extract maximum power from a PV module, while simultaneously forcing it to work at voltage close to the MPP, thus making it more efficient.

In this chapter, the MPPT technique was described using a deep architecture based on an long short-term memory (LSTM) network for solar PV $P_{\max}$ and $V_{\max}$ forecasting. By using the forecasting results, the MPPT technique can control and maintain reference voltage on a real-time basis. During one switching period, the PV module voltage V_{pv} and power $P_{\max}$ are checked, and a DC/DC converter is driven to operate at the voltage $V_{\max}$ corresponding to the actual MPP. MPPT units produce pulses that trigger the MOSFET switch in the DC-DC boost converter. Therefore, the DC-DC converter's power flow is controlled by varying the switching period's on/off duty cycle. Data-driven problems have been a major focus of machine learning techniques in recent decades. Machine learning techniques encompass a wide range of topics, such as statistics, mathematics, artificial neural networks, data mining, optimization, and artificial intelligence. Machine learning techniques try to identify relationships between input and output data, regardless of whether it is mathematically based. Once the machine learning models are properly trained with training data, decision-makers can use the forecasting input values to produce accurate forecasting results. In machine learning, data preprocessing plays a crucial role and can lead to enhanced performance [22]. Typically, machine learning uses three types of learning methods: supervised learning, unsupervised learning, and reinforcement learning. In the phase of training, supervised learning takes advantage of labeled data. For training data that has not been labeled in advance, unsupervised learning involves automatically classifying input data into clusters based on certain criteria. As such, clustering criteria determine how many clusters are created. Reinforcement learning's purpose is to obtain feedback from the external environment so as to maximize expected benefits. Numerous theoretical mechanisms and applications are proposed in accordance with three basic principles [23]. Due to rapid technological advancements in hardware and software, deep learning, a subfield of machine learning, has seen tremendous growth in the past few years. A wide range of applications have used deep learning to achieve satisfying results as it enables the realization of non-linear attributes and invariant data configurations at a high level [24]. A single machine learning model is also able to forecast renewable energy [25]. However, utilizing a single machine learning model cannot achieve better results owing to the difference in datasets, time steps, prediction ranges, settings, and performance indicators. Consequently, some studies focused on developing hybrid machine learning models or overall prediction methods for renewable energy predictions to improve their performance. The field of machine learning has been increasingly focusing on support vector machines and deep learning methods [26].

Solar radiation forecasting problems depend on previous time steps. These problems have been successfully addressed by deep learning [27, 28]. The recurrent neural network (RNN), i.e., LSTMs and gated recurrent units (GRUs), features that allow them to learn long-term dependencies are, therefore, more suitable for these problems [29, 30]. Deep learning based forecasting is mainly used for short-term forecasting within 1 or 2 hours to a few days [31, 32]. Solar irradiance forecasts for a period of 120 minutes are made in [33]. LSTM was used to forecast solar radiation for the upcoming year [34]. These methods did not account for maximum PV system power influence because they were trained with a supervised method without optimal feature extraction. Compared to conventional power production technologies, PV systems have become increasingly economically competitive over the past few years. Thus, they can be deployed as standalone units and in grid-connected applications, including in conjunction with electrical energy storage devices [35].

This book chapter is organized as the following sections: Related Literature, Background and Motivation, Overview of Machine Learning, Deep Learning Architecture, LSTM Network, Concepts of Solar PV and its MPPT Technique, Discussion of Simulation Results, and Conclusion.

6.2 Overview of Machine Learning

In machine learning, deep learning is a budding technique of soft computing. In the 1950s, artificial intelligence was implemented on hardware systems for the very first time. Within the 1960s, the concept of machine learning emerged with several systematic theorems. In addition, its recently evolved branch, deep learning, was first mentioned in the 2000s, and shortly afterward, its rapid application to numerous fields as a result of its new prediction concept derived from big data. The difference between deep learning with machine learning and artificial intelligence is shown in Figure 6.1.

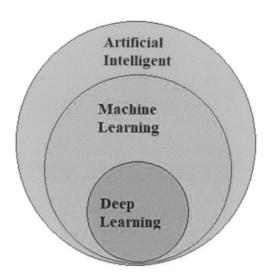

Figure 6.1 Difference between deep learning with machine learning and artificial intelligence.

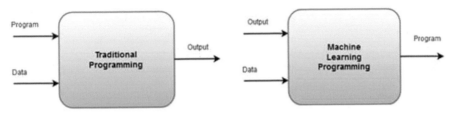

Figure 6.2 Difference between model (a) development of traditional programming and (b) development of machine learning programming.

Before getting into machine learning, we should understand the difference between humans and machines. If we ask "who is your father?" to a baby and a machine, the baby can recognize, but it cannot answer. But the machine cannot recognize or answer. This is because humans can learn from parents, friends, relatives, teachers, colleagues, society, books, and experience. So we should make the machine learn through learning algorithms to answer the above question. For that, we should feed problem content, data, and algorithms to the machine. In traditional programming, we generally provide programs and data as input to the computer/machine to get the desired2 output. But in machine learning programming, we should provide data and output to the computer to get the program as output. This can be depicted in Figure 6.2.

In general, machine learning technology combines three methods of learning: supervised learning, unsupervised learning, and reinforcement learning. Labeled information is beneficial to supervised learning during the training phase. To perform an unsupervised learning task, input data is spontaneously assigned to certain clusters based on certain criteria for training data that has not been labeled yet. In reinforcement learning, the external environment uses feedback to maximize expected advantages. The development of machine learning is depicted in Figure 6.3.

The first and one among the foremost crucial things to seek out in machine learning is to collect sufficient data and find what are the inputs and also the expected outputs. It is vital to keep in mind that machine learning can solely be utilized to learn patterns that exist in the training data; so it can identify what we have got before. So split the collected data into training as well as testing data. Once using machine learning, we tend to assume that the future will behave just like the past, and this is not invariably true. For this purpose, we want to divide our information into two parts: a training set that is used to learn the model; a test set to calculate the generalization performance of the

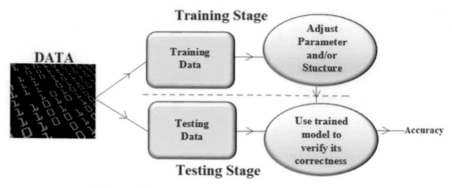

Figure 6.3   Development of machine learning model.

model. Then model assessment metrics are vital to enumerate the model. The model assessment aims to evaluate the generalization accuracy of a model on future information. Deep learning, a promising kind of machine learning able to predict non-linear features, has been reported in the literature. Its architectures are reported in the next section.

6.3 Deep Learning Architecture

Deep learning algorithm utilizes machine learning to achieve artificial intelligence. It has also shown hopeful outcomes in the unmanned vehicle, computer vision, audio processing, and data mining. Deep learning algorithms imitate human-level intelligence in machines to solve any problems for a given region. Deep learning architecture includes feed-forward neural network, RNN, and RNN extensions, i.e., LSTM and GRU.

A. Feed-Forward Deep Neural Network

A feed-forward deep neural network (FFNN) is the uncomplicated type of artificial neural network. Figure 6.4 depicts a simple example of an FFNN where i represents input, and h and o represent hidden and output, respectively. These neurons are interconnected. There are three sorts of layers referred to as the input, hidden, and output layers. In the hidden layer, the input is fed forward until the output is determined through the use of activation functions in each node and initialized weights and biases. At last, weights are changed utilizing the back-propagation algorithm and the loss functions in order to get the best results. In FFNN, there is no feedback

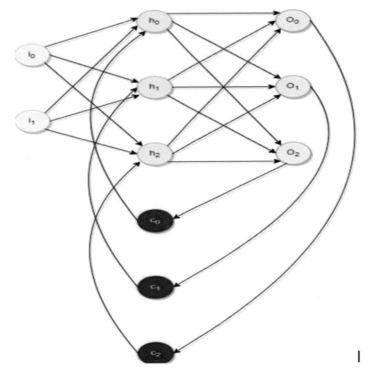

Figure 6.4 Simple FFNN.

connection; they do not have an instrument to recollect past outputs, unlike the RNN. Due to this, they are unsuitable for forecasting time series.

B. Recurrent Neural Network
RNN is an associate extension of a conventional FFNN where the output from the last step is taken as input to the current step. In other words, each node test will be based on the time step output from the previous test. An RNN is depicted in Figure 6.5.

The hidden state *ht* is defined as follows:

$$a(t) = b + Wh(t-1) + Ux(t);)); h(t) = \tanh(a(t)) \qquad (6.1)$$

$$o(t) = c + Vh(t); y(t) = soft\max(o(t)) \qquad (6.2)$$

where

$x(t)$ is taken as the input to the network at time step t;
$h(t)$ represents a hidden state at time t;

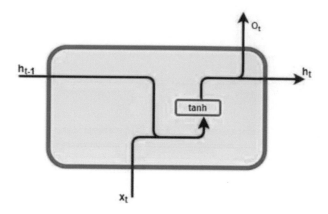

Figure 6.5   Simple RNN.

$o(t)$ illustrates the output of the network;
$y(t)$ is true target;
U is input to hidden connections parameterized by a weight matrix;
W is hidden-to-hidden recurrent connections parameterized by a weight matrix;
V is hidden-to-output connections parameterized by a weight matrix.

After the hidden state is obtained, in a fully connected layer, RNN output $o(t)$ is directly calculated from the current hidden state.

Unfortunately, it is a troublesome task to train an RNN as well as it cannot process very long sequences when tanh or relu is employed as an activation function. It is additionally determined that its gradient vanishes and tends to explode throughout training RNN to capture long-term dependencies. To handle the aforementioned problems, LSTM unit was introduced, and, afterward, GRUs were introduced recently.

C. Long Short-Term Memory
LSTM captures long-term dependencies to solve the gradient problems. Three gates are present in an LSTM unit. They are input gate, output gate, and forget gate with internal memory and is shown in Figure 6.6. An LSTM unit is ready to decide whether or not to retain or forget the current memory via the introduced gates. LSTMs use their input gate to decide which current data to pass. The forget gate determines what information has to be passed. The output gate specifies the state information to be passed.

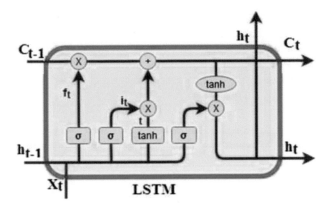

Figure 6.6   LSTM unit.

D. Gated Recurrent Units

LSTM is similar to GRU as it memorizes important information as well as transports it over long distances, capturing long-term relationships. In addition to being less complex, it is also more efficient in terms of computation. Certain data shows that it performs faster and better than LSTM. A GRU also features two gates, namely a reset gate and an update gate. Figure 6.7 illustrates how the above gates modulate the flow of data within the unit without having distinct memory cells.

GRU reset gate is defined as follows:

$$z = \sigma(W_z.x_t + U_z.h(t-1) + b_z);; r = \sigma(W_r.x_t + U_r.h(t-1) + b_r) \quad (6.3)$$

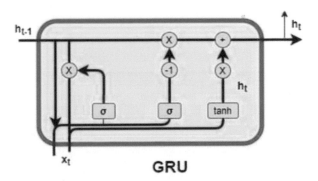

Figure 6.7   GRU unit.

$$h = \tanh(W_h.x_t + r*U_h.h(t-1) + b_z;; h = z*h(t-1) + (1-z)*h \quad (6.4)$$

where

    r is the relevance gate;

    h is the current cell state;

    $h(t - 1)$ is the previous cell state;

    W, U are weights;

    z is the update gate.

6.4 LSTM Network Based Prediction

In situations where the order of information is crucial, normal neural networks do poorly. To overcome this constraint, RNNs were developed. An RNN cell does not just consider its current input but also the output of RNN cells preceding it for its current output. RNNs are smart at managing sequential data; however, they run into problems when the context is much away. LSTM cell is a special type of RNN that is designed to overcome the problems associated with RNN in learning long-range dependencies. This is a new kind of neuron designed to overcome the limitations associated with RNN in learning long-term dependencies. As shown in Figure 6.8, LSTM networks are formed by repeating modules of different structures. As shown in Figure 6.8, LSTM networks are formed by repeating modules of different structures. In addition, LSTM consists of internal state data which is transmitted from one cell to another and also modified by operation gates. Additionally, they contain a sigmoid neural network layer and a point-wise

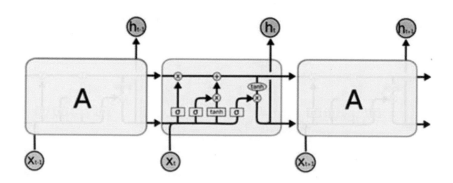

Figure 6.8 LSTM network model.

multiplication step. The sigmoid type layer delivers output as a number which is between 0 and 1. That output number describes what quantity of every part ought to be let through. The output value of 0 suggests that "let nothing through," whereas a value of 1 suggests that "let everything through." An LSTM has three of those gates to save and control the cell states. They are forget, input, and output gates. Every LSTM module comprises four layers or structures. The primary layer of LSTM is to come to a decision on what data is suitable to throw out from the past states (h_{t-1} and x_t) and is set by sigmoid or forget gate layer based on output C_{t-1}. If C_{t-1} is 1, it denotes "completely keep" and else "completely get rid of" which is shown in the following equation:

$$f_t = \sigma(W_f.[h_{t-1}, x_t] + b_f).$$ (6.5)

The next step is to resolve what new data is going to store in the state. This has another two layers whose function is shown in the following equations:

$$i_t = \sigma(W_i.[h_{t-1}, x_t] + b_i)$$ (6.6)

$$C_t = \tanh(W_C.[h_{t-1}, x_t] + b_C).$$ (6.7)

The first layer is called the sigmoid layer or the input gate layer selects which value is going to be updated. Next tanh layer produces a vector of the latest values which is able to be added to the state shown in the following equation:

$$C_t = \tilde{f}_t * C_{t-1} + i_t * C_t.$$ (6.8)

Creating and updating the state is done with the third layer, which combines the two layers above. The fourth or final state is to update the previous state C_{t-1} into the latest current state C_t. At last, multiplication is performed on the previous state by f_t, after forgetting then to add. Equation (6.7) is that the new value decided to update every state value. In the future, the output will be based on a filtered version of the cell state. There can be a sigmoid layer that determines what parts of the state are going to be the output. To place tanh in between −1 and 1, multiply the output of sigmoid gate by the state, and because of that, the only output that is determined is given in the following equations:

$$O_t = \sigma(W_O[h_{t-1}, x_t] + b_o)$$ (6.9)

$$h_t = O_t * \tanh(C_t)$$ (6.10)

where
> f_t is the forget gate;
> i_t is the update gate;
> C_t is the current cell state;
> h_{t-1} is the previous output;
> x_t is the input;
> W is the weight;
> O_t is the output gate.

6.5 Concepts of Solar PV and its MPPT Techniques

The solar PV systems transform solar daylight directly into electricity. However, economical utilization of PV systems is achieved through MPPT, as this guarantees maximum possible power transport to the grid. At a single point on its current-voltage (I–V) or power-voltage (P–V) characteristics curve of a PV cell, called MPP, the PV system functions with good efficiency and delivers high output power. There are various types of MPPT algorithms projected in the literature. The conventional MPPT techniques for example perturb and observe (P&O) is a simple, easily implemented, and efficient method. The PV module provides an output that is directly related to the irradiation and is inversely related to its temperature. The concepts behind the P&O MPPT method are given below. If $dP/dt > 0$, the perturbation should be in one direction and if $dP/dt < 0$, the perturbation should be reversed. The process should be repeated periodically until $dP/dt = 0$ reaches the maximum power point (MPP). This perturbation MPPT demands an extended response time to attain the MPP and more oscillation around MPP, as they do not have a previous value of the MPP at the beginning of the perturbation. This P&O algorithm is commenced from some random low value of voltage or some percentage of its open-circuit voltage and progresses via successive perturbations to attain its MPP in [10]. To achieve this approach, the range of increment or decrement value is crucial. Large ranges result in the algorithms obtaining MPP quickly but in oscillations around it. In contrast, a small range results in little oscillation around the MPP; however, the rate of convergence is also reduced. To overcome this problem, variable step ranges are used. To accomplish this, complicated control technology is needed. Less oscillation and quick tracking can be carried out in P&O with previous knowledge of reference voltage value. For the P&O MPPT, a deep learning based LSTM network predicted V_{max} is projected to offer a reference voltage V_{ref} online.

6.6 Simulation Results and Discussion

Figure 6.9 depicts the overall circuit design. It consists of a cascaded (cascaded buck and boost) converter with a 300-Wp PV array similar to real PV system specification (given in Table 6.1), P&O MPPT controller, blocking diode, and resistive loads. The simulation was done for a 1000-Ω resistive load, the inductor values were chosen as 564 μH and 0.78 H, and output capacitor value is taken to be 22 μF and 220 μF. P&O MPPT controller gets the starting value of perturbation derived from deep learning based LSTM network prediction.

6.6.1 Modeling and Performance Analysis

A. Solar PV System

PV cells are modeled by evaluating the *I–V* and *P–V* curves to simulate the real PV cell of varying atmospheric values. The most popular concept is to employ the single-diode model of the electrical equivalent circuit. This model comprises three parameters like open-circuit voltage, short circuit current,

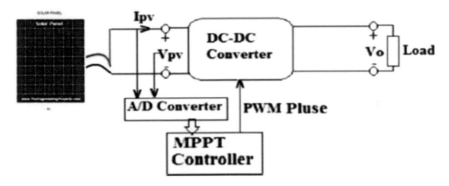

Figure 6.9 Overall system developed in MATLAB software.

Table 6.1 PV array specification.

S. No.	Parameters	Rating
1.	Maximum power (*P*max)	300 Wp
2.	Open-circuit voltage (*V*ocn)	36.1 V
3.	Short circuit current (*I*scn)	8.3 A
4.	Maximum power point voltage (*V*pm)	44.6 V
5.	Maximum power point current (*I*mp)	8.87 A

diode identity factor, as well as the modified one that includes a series resistor. Figure 6.10 shows a PV cell with a single diode.

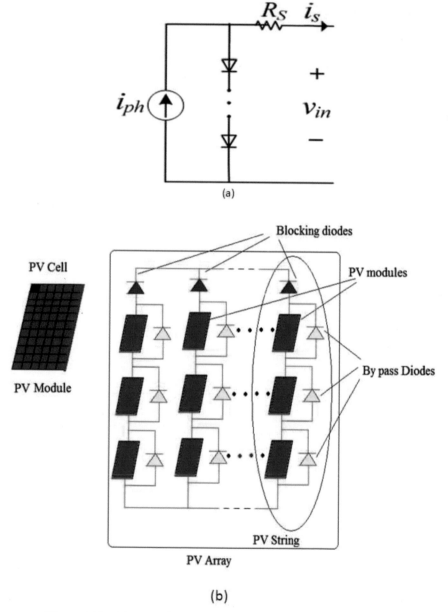

Figure 6.10 (a) PV cell single-diode model. (b) PV array configuration.

In a PV array, PV cells are connected in series and parallel. Arrays with parallel connections usually have an increased current, while modules with series connections have an increased voltage. Solar cells can typically be modeled as a current source connected to an inverted diode parallel to it. It has its own series and parallel resistance as shown in Figure 6.10.

$$i = I_{\text{ph}-}I_{s(T)}(\exp\{q(V_{\text{in}} + i.R_s)/KTA\} - 1 \tag{6.11}$$

where i is the PV panel output current, I_{ph} is photocurrent, $I_s(T)$ is the reverse saturation current, q (= 1.6×10^{-19}) is an electron charge, V_{in} is the terminal voltage of the PV panel, R_{SH} is the PV panel series resistance, A is the ideal factor of the PN junction of the PV diode, which varies in the range of [1, 2], and k (= 1.38×10^{-23} J/K) is the Boltzmann constant. The photocurrent is then found using the following equation:

$$I_{\text{ph}} = [I_{\text{sc}} + K_i(T - T_{\text{ref}})].G \tag{6.12}$$

where I_{sc} is the short circuit current provided by the PV panel at a reference temperature and an irradiance of 1 kW/m^2, K_i(= 3 mA/$^\circ$C) is the temperature coefficient, G is the solar irradiance in kW/m^2, and T and T_{ref} are measured temperature and reference temperature, respectively. The output current is then given in the following equation:

$$I_{\text{st}}(T) = I_s(T_{\text{ref}})\exp\{K_s(T - T_{\text{ref}})\} \tag{6.13}$$

where $i_s(T_{\text{ref}})$ is the reverse saturation current (T_{ref} = 295 K) and $K_s(\approx 0.072/^\circ C)$ is the temperature coefficient of the PV panel.

Figure 6.11(a) shows the I–V characteristics of a typical solar cell. Multiplying the voltage and current characteristics gives us the P–V characteristic shown in Figure 6.11(b). At MPP, the panel power output is at its maximum. Table 6.2 lists the standard PV array specifications based on NAVISOL datasheets.

The PV cell and array are modeled using a single-diode equivalent circuit and its performance characteristics are analyzed under varying atmospheric as shown in Figures 6.12(a) and (b).

B. Cascaded Converter Modeling

A design of cascaded buck and boost converters has been proposed in this work and whose circuit diagram is depicted in Figure 6.13.

The design equations consist of buck and boost converters combined with effects of inductor coupling on key converter performance parameters

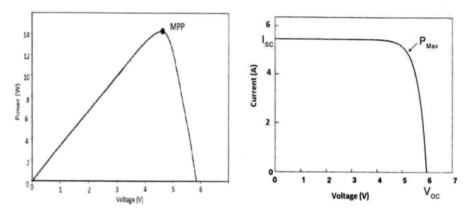

Figure 6.11 (a) *I–V* characteristics. (b) *P–V* characteristics.

Table 6.2 PV array specification.

S. No.	Parameters	Rating
1.	Maximum power (P_{max})	300 Wp
2.	Open-circuit voltage (V_{ocn})	44.6 V
3.	Short circuit current (I_{scn})	8.87 A
4.	Voltage at maximum power point (V_{pm})	36.1 V
5.	Current at maximum power point (I_{mp})	8.3 A

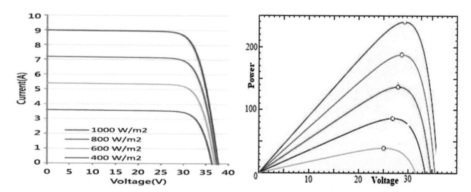

Figure 6.12 (a) *V–I* curve. (b) *P–V* curve under varying atmosphere condition.

like inductor ripple current, input ripple current, and minimum load. The following sections summarize the design of the DC-DC buck and boost converters formula and the values are tabulated in Table 6.3.

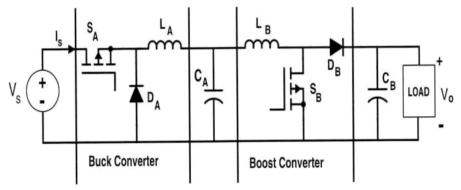

Figure 6.13 Circuit diagram of cascaded DC–DC converter.

Table 6.3 Cascaded buck and boost converter parameters.

S. No.	Parameters	Theoretical values
1.	Input voltage V_{in}(avg)	9.5 V
2.	The efficiency of the converter (η)	99%
3.	Buck Inductor (L_1)	0.78 H
4.	Boost_Inductor (L_2)	564 μH
5.	Buck_Capacitor (C_1)	220 μF
6.	Boost_Capacitor (C_2)	22 μF
7.	Switching frequency (F_S)	25 kHz
8.	Buck_Duty cycle ($D1$)	42.10%
9.	Boost_Duty cycle ($D2$)	20.83%
10.	Load resistance (R_L)	1000Ω
11.	(ΔVout)(ΔVout) Output voltage ripple	1.63 V
12.	Inductor ripple current (ΔI_L)	1.350 A
13.	Output voltage (V_{out})	12 V (boost)
14.	Output voltage (V_{out})	V (buck)

1. **Input voltage ranges:**
   Input voltage V_{in} (min)   = 8.0 V
   Input voltage V_{in}(max)   = 11.0 V
   Input voltage V_{in}(avg)   = 9.5 V
2. **Output voltage and current ranges:**
   Output voltage V_o (boost)   = 12.0 V
   Output voltage V_o(buck)   = 4.0 V
   Output current I_o = 1.0 A

3. **Calculation of duty cycle (boost):**

   Duty cycle $D1 = $ $= 20.83\%$ $D1 = 1 - \dfrac{Vin(avg)}{Vout}$

4. **Calculation of duty cycle (buck):**

   Duty cycle $D2 = $ $= 42.10\%$ $D2 = \dfrac{Vout}{Vin(avg)}$

5. **Calculation of inductor (boost):**

   Inductance $L1 = $ $= 564\ \mu H$ $L1 = \dfrac{D1(Vout - Vin(avg))}{Fs * \Delta IL1}$

6. **Calculation of inductor (buck):**

   Inductance $L2 = $ $= 0.78\ H$ $L2 = \dfrac{D2(Vout - Vin(avg))}{Fs * \Delta IL2}$

7. **Calculation of inductor ripples current (boost):**
   Inductor ripple current ΔI_{L1} $= 20\%\text{-}40\%$ of $I_o = 0.03$ A

8. **Calculation of inductor ripples current (buck):**
   Inductor ripple current ΔI_{L2} $= 20\%\text{--}40\%$ of I_o
                    $= 0.118$ mA

9. **Output capacitor selection (boost):**

   Output capacitance $C_O = \left[\dfrac{I_0.D_1}{FS.\Delta Vout}\right] = 22 \mu F$

   where

   Output voltage ripple $\Delta V_{out} = \text{ESR}\left[\dfrac{I_0}{1 - D1} + \dfrac{\Delta IL}{2}\right] = 0.378$ V

10. **Output capacitor selection (buck):**

    Output capacitance $C_O = \left[\dfrac{\Delta IL2}{8 * FS.\Delta Vout}\right] = 220\ \mu F$

    where

    Output voltage ripple $\Delta V_{out} = \text{ESR}\left[\dfrac{I_0}{1 - D2} + \dfrac{\Delta IL}{2}\right] = 0.00268$ mV.

C. Proposed MPPT Controller

To obtain the maximum power from a solar array, MPPT algorithms are required. The MPPT of solar panels varies with the amount of irradiation and temperature; so MPPT algorithms are necessary. Figure 6.14 illustrates a modified P&O algorithm flowchart.

A comparison of the proposed and conventional algorithms for power-conditioning of PV systems is shown in Figure 6.15. Both algorithms respond differently to the same variation in irradiation. The proposed algorithm took 0.2 seconds lesser to reach the steady-state than the conventional P&O algorithm.

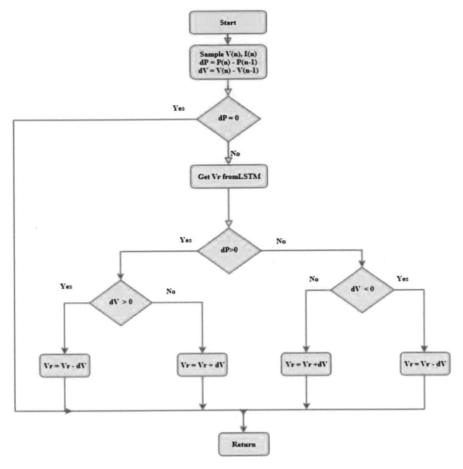

Figure 6.14 Flowchart of modified P&O MPPT algorithm.

6.6.2 Prediction or Forecasting Methodology

Time series analysis denotes the analysis of variation of data over a period of time. One of the applications of time series analysis is solar forecasting. In this section, time series analysis is performed using an LSTM network to predict maximum voltage (V_{mpp}) of 300 Wp real-time solar PV system based on its MPP of the past one year. The data used in this work is collected from 29th September 2018 to 29th September 2019. For the purpose of training, MPP data from 29th September 2018 to 29th July 2019 is taken (75%). For the sake of prediction, MPP data from 1st August 2019 to 29th September

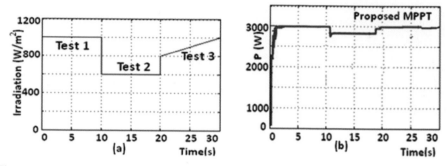

Figure 6.15   (a) Influence of irradiation variation. (b) Influence of temperature variation. (c) Response of the proposed MPPT algorithm in PV system.

2019 is utilized (25%). To estimate the performance of the proposed algorithm compared with the real MPP for the same month data as well. The database contains ten columns: date, time, voltage, ampere, kWh, PF, kW, P_{max}, V_{max}, and hours. The predicting value is V_{max}; therefore, there is no need to give any interest to the rest of the column data.

Figure 6.16 shows the recorded value of V_{max} collected from the data logger of real solar 300-Wp PV system for the past one-year day-wise 365 counts. It is found to be extremely non-linear and it is very difficult to capture the trend using this information. Therefore, that LSTM is utilized in this work. LSTM is a type of RNN able to recall the previous information, and while predicting the upcoming values, it captures this previous information into account.

The steps involved in LSTM network development are the same as any other machine learning problem. The aim of this chapter work is to predict V_{max} and to provide a reference voltage for the P&O MPPT method. LSTM network training, LSTM network testing, and sequence data loading are the steps required for time series forecasting utilizing deep learning. The first step is to import the dataset and normalize/scale our data between 0 and 1 using MinMaxScaler. The succeeding step is to divide the data into training as well as testing data. Among 364 data, first 300 data are divided as training data and the remaining 64 as testing data. The fourth step is to create an LSTM model. Figure 6.17 shows LSTM network based test and predicted (scaled) value of V_{max}. In the output waveform, the actual V_{max} (scaled) value for the July month represents the blue line and August 2019, while the red line presents the predicted V_{max} value. It is clearly shown that the predicted value follows the tested value.

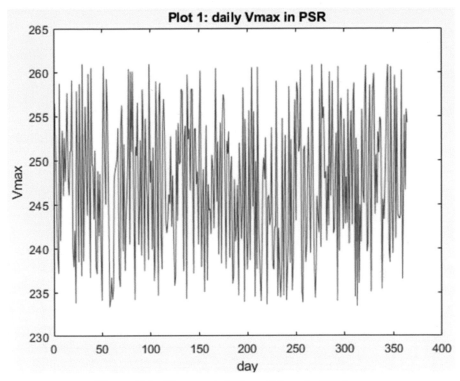

Figure 6.16   The recorded value of V_{max} for 365 days.

The comparison has also been made for the recorded V_{max} from the data logger and LSTM algorithm predicted value and is tabulated in Table 6.4. It is also observed that the error percentage in both cases is very less in the proposed LSTM algorithm. The training progress of V_{max} is also shown in Figure 6.18.

6.6.3 Utilizing Predicted Value in MPPT Technique

In the P&O-based algorithm, MPP may be attained by comparing the power difference dP. LSTM network training, LSTM network testing, and sequence data loading are the steps required for time series forecasting utilizing deep learning. This voltage reference (V_{ref}) can either be maximized or minimized with a small value that is forced to operate the PV array. The most appropriate way to use the conventional P&O MPPT method is to have a previous understanding of the V_{ref} value, which will diminish the tracking period of the MPPT method. For that, we can make use of deep learning based LSTM

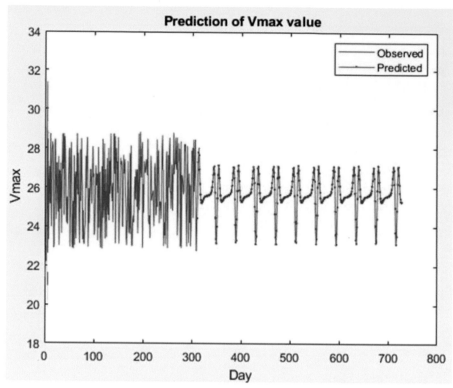

Figure 6.17   The predicted value of V_{max} for another 365 days.

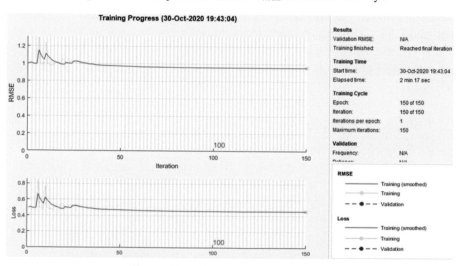

Figure 6.18   Training progress of V_{max}.

Table **6.4** Comparison of recorded and predicted V$_{max}$.

Day	Date	Recorded value	Predicted value	Error percentage (%)
351st	29.07.2019	28.8033	24.03	−8.26974
352nd	30.07.2019	24.0425	26.21	3.494287
353rd	31.07.2019	28.078	27.13	−8.49543
354th	01.08.2019	27.0213	29.53	−8.4984
355th	02.08.2010	22.8388	24.96	−10.311
356th	03.08.2019	21.2563	23.7	−1.98981
357th	04.08.2019	25.0024	25.51	−8.26974

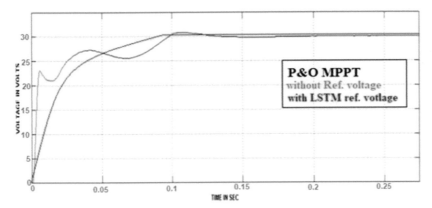

Figure 6.19 Converter output voltage with and without the utilization of LSTM V_{ref} value in P&O MPPT.

network forecasting parameters as the reference values. Figure 6.14 depicts the flowchart of P&O MPPT which utilized day-wise LSTM predicted V_{max} as V_{ref}.

Figure 6.19 inferred that the P&O MPPT with reference value oscillates less and reaches its maximum value 0.15 seconds earlier than the other case.

6.7 Conclusion and Future Directions

This chapter presents the modified P&O MPPT algorithm whose initial perturbation or reference value is provided by a deep learning based LSTM network to increase its tracking speed and response. Day-wise maximum voltage of 365 days is collected from 33-kW solar PV systems to predict maximum voltage which is further utilized for reference voltage or initial perturbation value for P&O MPPT. To determine how well the LSTM network predicts, the statistical error is calculated. It is found that LSTM is superior in prediction methodology. Ultimately, it can be inferred that the proposed P&O MPPT provides a good response, less tracking time, and very little oscillation around the MPP value. This section also briefly presents some interesting research directions, which are worth investigating further. GRU method can be implemented instead of LSTM unit in the same problem to accomplish more precise prediction. In near future, we can try to implement it in Arduino UNO hardware as an MPPT controller to evaluate the proposed methodology.

References

[1] Kotsiantis, S.B. Supervised machine learning: A review of classification techniques. Informatica **2007**, 31, 249–268.

[2] Qiu, J., Wu, Q., Ding, G., Xu, Y., Feng, S. A survey of machine learning for big data processing. EURASIP J. Adv. Signal Process. **2016**, 2016, 67.

[3] Gu, J., Wang, Z., Kuen, J., Ma, L., Shahroudy, A., Shuai, B., Liu, T., Wang, X., Wang, L., Wang, G., *et al*. Recent advances in convolutional neural networks. Pattern Recognit. **2017**, 1, 1–24.

[4] Amasyali, K., El-Gohary, N.M. A review of data-driven building energy consumption prediction studies. Renew. Sustain. Energy Rev. **2018**, 81, 1192–1205.

[5] Wang, H.Z., Lei, Z.X., Zhang, X. A review of deep learning for renewable energy forecasting. Energy Convers. Manage. **2019**, 198, 111799.

[6] Olabi, A.G. Renewable and energy storage system. Energy **2017**, 136, 1–6.

[7] Zendehboudi, A., Baseer, M.A., Saidur, R. Application of support vector machine models for forecasting solar and wind energy resources: A review. J. Clean. Prod. **2018**, 199, 272-285.

[8] Bosch, J., Kleissl, J. Cloud motion vectors from a network of ground sensors in a solar power plant. Sol. Energy **2013**, 95, 13–20.

[9] Dong, Z., Yang, D., Reindl, T., Walsh, W.M. Satellite image analysis and a hybrid ESSS/ANN model to forecast solar irradiance in the tropics. Energy Convers. Manage. **2014**, 79, 66–73.

[10] Hohm, D.P., Ropp, M., 2000. Comparative study of maximum power point tracking algorithms using an experimental, programmable, maximum power point tracking test bed, in: Conference Record of the Twenty-Eighth IEEE Photovoltaic Specialists Conference, 2000, pp.1699–1702.

[11] Inman, R.H., Pedro, H.T., Coimbra, C.F. Solar forecasting methods for renewable energy integration. Prog. Energy Combust. Sci. **2013**, 39, 535–576.

[12] Mahamadou, A.T., Mamadou, B.C., Brayima, D., Cristian, N. Ultracapacitors and batteries integration for power fluctuations mitigation in wind-PV-diesel hybrid system. Int. J. Renew. Energy Res. **2011**, 1, 86–95.

[13] Yang, D., Sharma, V., Ye, Z., Lim, L.I., Zhao, L., Aryaputera, A.W. Forecasting of global horizontal irradiance by exponential smoothing, using decompositions. Energy **2015**, ISSN 0360-5442.

[14] Teleke, S., Baran, M., Bhattacharya, S., Huang, A. Rule-based control of battery energy storage for dispatching intermittent renewable sources. IEEE Trans. Sustain. Energy **2010**, 1,117–124.

[15] Quesada-Ruiz, S., Chu, Y., Tovar-Pescador, J., Pedro, H., Coimbra, C. Cloud-tracking methodology for intra-hour {DNI}forecasting. Sol. Energy **2014**, 102, 267–275.

[16] Yang, H., Kurtz, B., Nguyen, D., Urquhart, B., Chow, C.W., Ghonima, M., Kleissl, J. Solar irradiance forecasting using a ground based sky imager developed at UC San Diego. Sol. Energy **2014**, 103, 502–524.

[17] Guerriero, P., Napoli, F.D., Vallone, G., d'Alessandro, V., Daliento, S. Monitoring and diagnostics of PV plants by a wireless self-powered sensor for individual panels. IEEE J. Photovolt. **2016**, 6, 286–294.

[18] Nikitidou, E., Kazantzidis, A., Tzoumanikas, P., Salamalikis, V., Bais, A. Retrieval of surface solar irradiance, based on satellite-derived cloud information, in Greece. Energy **2015**, 90(Part 1), 776–783.

[19] Yang, D., Quan, H., Disfani, V.R., Liu, L., Reconciling solar forecasts: Geographical hierarchy. Solar Energy **2017**, 146, 276–286.

[20] Wu, X., Zhu, X., Wu, G.Q., Ding, W. Data mining with big data. IEEE Trans. Knowl. Data Eng. **2014**, 26, 97–107.

[21] Wang H., Ruan J., Wang G., *et al*. Deep learning based interval state estimation of AC smart grids against sparse cyber attacks. IEEE Trans. Ind. Inf. **2018**, 14(11), 4766–4778.

[22] Benjamin, P., Marc, M., Philippe, P., Batista, D. J. Historical trends in global energy policy and renewable power system issues in Sub-Saharan Africa: The case of solar PV. Energy Policy **2019**, 127, 113–24.

[23] Yongning, Z., Lin, Y., Zhi, L., Xuri, S., Yansheng, L., Su, J. A novel bidirectional mechanism based on time series model for wind power forecasting. Applied Energy **2016**, 177, 793–803.

[24] Laura, F.-P., Fermín, M., Martín, G.-R., Teresa, L. Assessing energy forecasting inaccuracy by simultaneously considering temporal and absolute errors. Energy Convers Manage **2017**, 142, 533–546.

[25] Short, W., Packey, D.J., Holt, T.A. Manual for the Economic Evaluation of Energy Efficiency and Renewable Energy Technologies. Golden, CO, USA: U.S. Department of Energy Managed by Midwest Research Institute for the U.S. Department of Energy, 1995.

[26] Gasparin, A., Lukovic, S., Alippi, C. Deep learning for time series forecasting: The electric load case. arXiv, **2019**.

[27] Jafar, A., Lee, M. Performance Improvements of Deep Residual Convolutional Network with Hyperparameter Opimizations. Seoul, Korea: The Korea Institute of Information Scientists and Engineers, 2019, pp. 13–15.

[28] Sezer, O.B., Gudelek, M.U., Ozbayoglu, A.M. Financial time series forecasting with deep learning: A systematic literature review. 2005–2009, arXiv, **2019**.

[29] Bengio, Y., Simard, P., Frasconi, P. Learning long-term dependencies with gradient descent is difficult. IEEE Trans. Neural Netw. **1994**, 5, 157–166.

[30] Chung, J., Gulcehre, C., Cho, K., Bengio, Y. Empirical evaluation of gated recurrent neural networks consequence modeling. arXiv, **2014**.

[31] Aslam, M., Lee, J., Kim, H., Lee, S., Hong, S. Deep learning models for long-term solar radiation forecasting considering microgrid installation: A comparative study. Energies **2019**, 13, 147.

[32] Che, Y., Chen, L., Zheng, J., Yuan, L., Xiao, F. A novel hybrid model of WRF and clearness index-based Kalman filter for day-ahead solar radiation forecasting. Appl. Sci. **2019**, 9, 3967.

[33] Zhang, X., Wei, Z. A hybrid model based on principal component analysis, wavelet transform, and extreme learning machine optimized by bat algorithm for daily solar radiation forecasting. Sustainability **2019**, 11, 4138.

[34] Lai, C.S., McCulloch, M.D. Levelized cost of electricity for solar photovoltaic and electrical energy storage. Appl. Energy 2017, 190, 191–203.

[35] Huang, C., Wang, L., Lai, L. Data-driven short-term solar irradiance forecasting based on information of neighboring sites. IEEE Trans. Ind. Electron. **2019**, 66, 9918–9927.

7

Application of Optimization Technique in Modern Hybrid Power Systems

D. Lakshmi[1], R. Zahira[2], C. N. Ravi[3], P. Sivaraman[4], G. Ezhilarasi[5], and C. Sharmeela[6]

[1]Academy of Maritime Education and Training (AMET), India
[2]BSA Crescent Institute of Science and Technology, India
[3]Vidya Jyothi Institute of Technology, India
[4]Vestas Technology R&D Chennai Pvt Ltd, India
[5]Sri Sairam Institute of Technology, India
[6]Anna University, India
E-mail: lakshmiee@gmail.com; zahirajaved@gmail.com;
sivaramanp@ieee.org; sharmeela20@yahoo.com

Abstract

The load frequency control (LFC) is one of the operational problems in the power system. The supply of reliable and quality power to its users is the main aim of any electric power utility. The reliability of the power supply requires that the demand and losses should be equal to power generation in the power system. This balance is measured with the help of frequency in the output line. Hence, controlling the generation based on the change in load is carried out by the LFC. The frequency in the LFC areas and the tie line connecting the LFC areas should be the same to ensure reliability. In this research work, deregulated market structure two-area LFC and a tie line connecting them are considered. Intelligent algorithms provide better performance compared to the conventional optimization technique. For the LFC problem in the two-area power system, this paper considers the intelligent algorithms, namely flower pollination algorithm (FPA) and differential evolution (DE). The hybrid DE-FPA algorithm is being developed to manage the generating plant's generation based on load variations at the

149

critical damped time. Hybrid intelligent algorithms use the advantages of the DE and FPA and give better results as compared to conventional control techniques.

The need of the day is to use renewable energy sources, and, in particular, in this work, wind energy generation is considered. In wind energy generators, doubly fed induction generators (DFIGs) are commonly used. Controlling the electric power output of DFIG is complex and requires better control techniques. The main control variable of the DFIG is inertia, and it is controlled by the controller. Control of the frequency of generated power is controlled by the kinetic energy in the wind turbine blade. LFC consisting of DFIG is considered and controllers are designed to take care of the response of conventional generators and DFIG for the change in load demand. The results of the developed algorithm are compared with the conventional controller and discussed in the research work.

7.1 Introduction

The main purpose of power system dynamics is to maintain a balance between production and load. The proper quality of the power supply index requires that the frequency and voltage be maintained within acceptable limits. To improve system efficiency and meet demand, two or more regions are interconnected by a tie line. The load variation within an area affects the remaining areas, which will reflect the frequency [load frequency control (LFC) - actual power] and voltage (AVR - reactive power changes).

In this work, we are considering the real power output of the generating unit (LFC). LFC plays an important role in power system operation. The objective of LFC is to control the real power of the generation to meet the dynamic change in load demand. Thus, LFC regulates the MW power output of generators within a control area in response to the changes in system frequency and tic-line power, called area control error (ACE) [1].

In the literature, there are various types of control methods or logics/algorithms proposed by various researchers for LFC regulation. In the early stages integral, proportional–integral (PI), and proportional-integral derivative (PID) have been used. Among them, PI is mostly used because of its simplicity and is tuned using the Ziegler-Nichols method [2]. There are numerous artificial intelligence based controllers used for LFC regulation to reduce the settling time, undershoot, and overshoot. Some of the artificial intelligence based controllers are two-stage neural network fuzzy logic, neuro-fuzzy controllers, etc., and they are more suitable for modern complex power systems with nonlinear characteristics [3, 4].

In the modern era with new technologies in the power industry, the electrical power system is transferred to more emphasis on restructuring and deregulation. The deregulated power system splits the conventional single power system (from generating station to distribution) into three different sections/verticals, namely generation companies in short form GENCOs, transmission companies in short TRANSCOs, and distribution companies in short DISCOs. These three different verticals are managed by independent service or system operator (ISO). Various researchers have analyzed the LFC problem in a deregulated environment [5, 6]. The DISCO participation matrix (DPM) based LFC regulation for area control error participation factor (APF) in the deregulated electrical power systems was discussed in [7].

In general, electrical power system consists of energy generation from conventional as well as non-conventional energy resources, i.e., hydro, coal, nuclear, solar, wind, biomass diesel, etc.; nowadays, non-conventional energy resources are gradually replacing conventional sources because they are free from environmental pollution and emissions, zero or lesser running cost, etc. Wind energy is the domestic source of energy and is available in abundance. In a deregulated power system, LFC becomes complicated when wind units are integrated with the system.

Inertial control of the wind units provides coordination with conventional systems [8, 9]. A wind unit supports the system by inertial control, pitch control, and speed control [10]. Participation of doubly fed induction generator (DFIG) is analyzed through modified inertial control, which reflects the frequency deviation by making use of kinetic energy of the turbine blades to improve the frequency [11, 12].

Research on three areas deregulated power systems with optima PID using imperialist competitive algorithms was discussed in [13]. Dynamic participation of DFIG for hydrothermal deregulated power systems with fuzzy controllers was discussed in [14]. Participation of DFIG in an asynchronous power system using feedback control is discussed in [15] and the use of intelligent algorithms such as firefly for deregulated power systems in [16].

7.2 Modern Power System

The electric power industry has operated as a vertically integrated utility (VIU) for many years. VIU is self-sufficient in terms of power generation, transmission, and distribution. VIU owns all levels of the power supply circle, including generation, transmission, and distribution and so has the sole right to set the price of electric energy. As a result, VIU refers to the electric power industry that is monopolized by a single utility.

Furthermore, while such interconnection improves reliability, it comes with some drawbacks, including inefficient production, large losses, decaying infrastructure, and poor management. As a result, deregulation in the electric power industry is required, as is the creation of a separate autonomous company for generation, transmission, and distribution.

7.2.1 Deregulated Power System

Deregulation and market-based competition are taking place in the electric power industry. Retail power sales and the separation of power generation and related services are among the responsibilities of this transformation (http://nptel.ac.in/courses).

Deregulation of the power industry was motivated by several factors.

1. Electricity rates are lower as a result of competition.
2. Customers have a wide range of options, resulting in greater plans, reliability, and quality.
3. Service enhancement, cost reduction, and profit maximization.
4. Innovation in the operation of GENCOs and DISCOs to improve service, resulting in cost savings and profit maximization.

Deregulated power system components are shown in Figure 7.1.

7.2.2 Components of Deregulation

Several new entities have emerged in the power industry as a result of deregulation.

Figure 7.1 Components of deregulation.

- **GENCOs (Generating Companies):**

GENCOs are a group of one or more generating units that are sold as a whole through a single ownership company structure. As a result, GENCOs produce electricity and have the option to sell it to companies with which they have concluded sales agreements.

- **TRANSCOs (Transmission Companies):**

The delivery of power from GENCOs to DISCOs or retailers is the responsibility of TRANSCO. The TRANSCO maintains and operates the transmission system in certain geographical regions to assure the overall reliability of the electric power system.

- **DISCOs (Distribution Company):**

The electricity is distributed to customers through a DISCO. The DISCO purchases wholesale electric power from GENCOs directly or through spot markets and distributes it to end customers. The DISCO is responsible for maintaining the availability and reliability.

- **RESCOs (Retail Energy Service Company):**

A RESCO is a company that sells electricity. RESCO buys power from GENCOs and sells it directly to customers.

- **Customer:**

In deregulation, consumer has a variety of options, including bidding for power on the spot market or contracting directly from a DISCO or a GENCO.

- **ISO (Independent Service Operator):**

An ISO is an independent organization that is not involved in market trading. It is responsible for assuring the electricity system's dependability and security. The ISO encourages ancillary services such as frequency regulation, emergency reserve supply, and reactive power from other system organizations.

- **Ancillary services:**

Ancillary services can be defined as "the services associated with those activities on the deregulated power system that are necessary to support the transmission of power by maintaining reliable operation and ensuring the required degree of quality and safety." North American Electric Reliability Corporation (NERC) has suggested 12 ancillary services. Their performances are as follows:

- to manage generation and load balance (control of frequency);

- to ensure reactive power support and the voltage;
- to retain transmission reserves and generation;
- to be ready in the event of an emergency (i.e., control of stability and system restart).

LFC is one of the ancillary services provided to maintain real power balance. The LFC also maintains the net scheduled area interchange and keeps the system frequency deviation at zero. Thus, the performance of LFC is important and a few of the qualities do.

1. On a market-price basis, LFC can be traded as an ancillary service. Because of the flexibility of participation in LFC, both generation and loads will be more dynamically regulated.
2. The nature of tie-line transactions is very dynamic. To guarantee that tie-line transactions remain stable, this must be taken into account when designing LFC controllers.
3. LFC should be able to track the generators' participation, which is also dynamic because DISCOs have the opportunity to contract with different GENCOs.

7.2.3 Types of Transactions

There are three types of structures based on the transaction among DISCOs and GENCOs. They are (1) pool-co or charged transaction (contract between DISCO and GENCO of the same area), (2) bilateral transaction (DISCOs have a contract with a GENCO of another area), and (3) charged-cum-bilateral transaction [18]. In this chapter, we will consider only the bilateral transaction in detail.

7.2.3.1 Bilateral transactions

A DISCO can deal with any GENCO in the same area or another area. As a result, the idea of physical control area has been superseded with virtual control area (VCA). The GENCO and DISCO affiliates are encircled by VCA's contract, which has adjustable borders. Each DISCO is in charge of managing the tie-line power exchange with its neighbors by securing transmission and generation as needed in a bilateral transaction (Bevrani et al. 2004). The behavior of the power system before and after deregulation is shown in Figure 7.2.

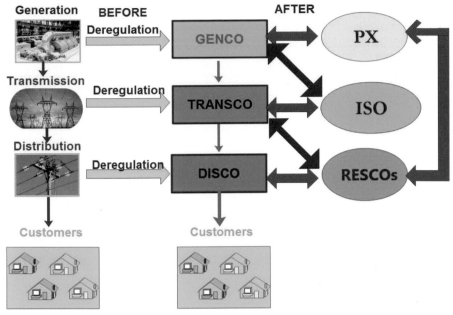

Figure 7.2   Structure of power system before and after deregulation.

7.2.3.2 DPM and APF

The DPM is used to depict the contracts that are formed between a DISCO and a GENCO [4]. Equation (7.1) shows the DPM for the power system, where "*ij*th" entry of the matrix (called generation participation factor) represents the fraction of the total contract by the DISCO "*j*" with a GENCO "*i*," and the sum of all the entries in a particular column is unity. Equation (7.1) gives the DPM matrix

$$\text{DPM} = \begin{bmatrix} \text{gpf}_{11} & \text{gpf}_{12} & \text{gpf}_{13} & \text{gpf}_{14} \\ \text{gpf}_{21} & \text{gpf}_{22} & \text{gpf}_{23} & \text{gpf}_{24} \\ \text{gpf}_{31} & \text{gpf}_{32} & \text{gpf}_{33} & \text{gpf}_{34} \\ \text{gpf}_{41} & \text{gpf}_{42} & \text{gpf}_{43} & \text{gpf}_{44} \end{bmatrix} \text{ and } \sum_{j=1}^{4}\sum_{i=1}^{4} \text{gpf}_{ij} = 1 \quad (7.1)$$

where $\text{gpf}_{ij} = \dfrac{\text{Demand of DISCO "}j\text{" from GENCO "}i\text{"}}{\text{Total Demand of DISCO "}j\text{"}}$.

The difference between an area's actual interchange and its scheduled interchange is known as ACE since there are several GENCOs in each area keeping the frequency at its specified value (ACE). Thus, control operations require ACE, which must be divided among them in accordance with their participation in the LFC. ACE participation factors (Apf) indicate the elements that provide ACE to the participating GENCOs, and the sum of the Apf in a given area must equal unity as shown in following equation:

$$\sum_{j=1}^{m} \text{Apf}_j = 1. \tag{7.2}$$

The generation of a GENCO in MW expressed in terms of gpfs is shown in the following equation:

$$\Delta P_G = \sum_{j=1}^{m} \text{gpf}_{ij} \Delta P_{Loj}. \tag{7.3}$$

Equation (7.4) shows the contract of DISCO with GENCO during the steady-state condition.

$$\Delta P_{Li,Loc} = \sum_{i=1}^{n} \Delta P_{Li}. \tag{7.4}$$

The scheduled steady-state tie-line power flow from area i to area j is given by the following equation:

$$\Delta P_{\text{tie1,2schedule}} = \sum_{i=1}^{n} \sum_{j \neq i}^{m} \text{gpf}_{ij} \Delta P_{Loj} - \sum_{i \neq j}^{m} \sum_{j=1}^{n} \text{gpf}_{ij} \Delta P_{Loj}. \tag{7.5}$$

In bilateral transactions, the tie-line power will not settle at zero, but rather at a value determined by a bilateral contract between GENCOs in one area and DISCOs in another.

7.2.4 Renewable Energy Sources

While harnessing nature's power is typically considered new technology, it has long been used for heating, transportation, lighting, and other purposes. The wind has propelled ships across the oceans and mills that process grain.

7.2.4.1 Doubly fed induction generator

Electrical power generation from renewable sources, such as wind, is drawing increased interest as a result of environmental concerns and a scarcity of traditional energy sources.

In general, wind turbines do not work with traditional energy sources. However, with advancements in controller technology, kinetic energy stored in wind turbines may be retrieved using variable speed generators. Wind turbines based on DFIG can run at varying speeds and synchronize with the frequency regulation of conventional energy sources. A dynamic model of the DFIG wind unit is shown in Figure 7.2.

In the emulation control of DFIG, a control signal is provided to modify the power setpoints ΔP_f^*, which is a function of deviation and rate of change of frequency. To get the most power, the controller will aim to keep the turbine at its ideal speed. ΔP_w^*. The controller provides the set power depending on the measured speed and measured electrical power. From eqn (7.6), ΔP_{NC} has two components they are additional reference points based on frequency change as shown in eqn (7.7) and the PI controller calculates the optimum turbine speed as a function of wind speed:

$$\Delta P_{NC} = \Delta P_f^* + \Delta P_\omega^* \tag{7.6}$$

$$\Delta P_f^* = \frac{1}{R_w} \Delta X_2. \tag{7.7}$$

Contribution of DFIG toward system inertia is given by the following equation:

$$\frac{2H}{f} \frac{d\Delta f}{dt} = \Delta P_G + \Delta P_{NC} - \Delta P_{tie1,2} - \Delta P_D - D\Delta f. \tag{7.8}$$

From Figure 7.3, it is seen that additional reference power setting based on the change in frequency using a washout filter with a time constant T_w and

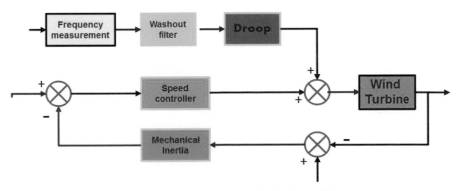

Figure 7.3 DFIG-based wind unit model.

R_w the droop constant and ΔX_2 measured change in frequency to which the wind system is connected.

DFIG wind unit reacts to frequency deviations during transients by using the stored kinetic energy and does not provide permanent system frequency deviation. DFIG inertia contributes to the system, and the real power injected by the wind unit is ΔP_{NC}. This injected power is compared with ΔP_{NCref} to obtain maximum power output.

7.2.4.2 DFIG in deregulated power system

Overall transfer modeling of deregulated power systems is shown in Figure 7.4. It shows the transfer function model of the system, in which R_1, R_2, R_3, and R_4 are the governor regulation parameters of thermal units for area 1 and area 2 in Hz/MW, respectively. Each area consists of a two-speed governing system and two reheat turbines. The transfer function model of the speed governor is given by the following equation:

$$G_{Gj}(s) = \frac{1}{1 + sT_{Gj}} \tag{7.9}$$

where T_{Gj} is the time constant of the jth governor. The transfer function model of the steam turbine is given by the following equation:

$$G_T(s) = \frac{1}{1 + sT_{Tj}} \tag{7.10}$$

where T_{Tj} is the time constant of the jth reheat turbine, K_{Rj} is the gain constant of reheater, and T_{Rj} is the time constant of the reheater. The transfer function of power system is as shown in the following equation:

$$G_{Pi}(s) = \frac{K_{Pi}}{1 + sT_{Pi}} \tag{7.11}$$

where K_{Pi} is the gain of the ith area power system and T_{Pi} is the time constant of the ith area power system. The values of K_P and T_P are given by the following equations:

$$K_{Pi} = \frac{1}{D_i} \tag{7.12}$$

$$T_{Pi} = \frac{2H}{f D_i} \tag{7.13}$$

where H is the per unit inertia constant, f is system frequency, and D_i is expressed as percent change in load by percent change in frequency.

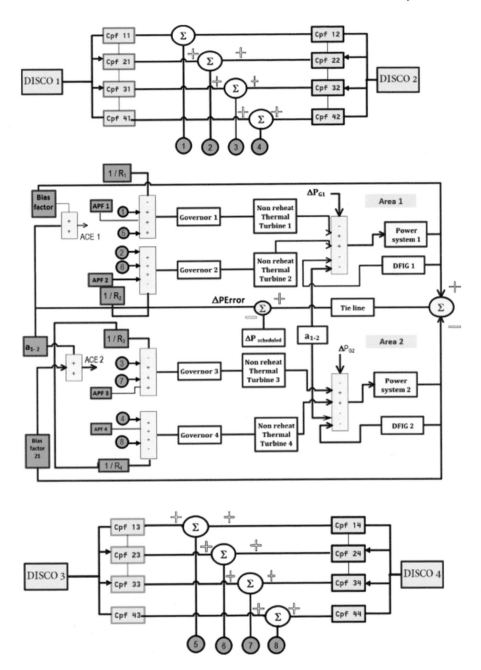

Figure 7.4 Transfer function model of two areas' deregulated power system.

The scheduled steady-state tie-line power flow from area 1 to area 2 is the difference between the demand of DISCOs in area 2 from GENCOs in the area 1 and the demand of DISCOs in area 1 from the GENCOs in area 2, which is given in the following equation:

$$\Delta P_{\text{tie1,2schedule}} = \sum\nolimits_{i=1}^{2} \sum\nolimits_{j=3}^{4} \text{gpf}_{ij} \Delta P_{Ljo} - \sum\nolimits_{i=3}^{4} \sum\nolimits_{j=1}^{2} \text{gpf}_{ij} \Delta P_{Loj}.$$
(7.14)

The tie-line power flow from area 1 to area 2 is the product of the tie-line coefficient and the difference between the change in frequency in area 1 and the change in frequency in area 2, as given by the following equation:

$$\Delta P_{\text{tie1,2actual}} = \frac{2\Pi T_{12}}{s} [\Delta f_1 - \Delta f_2].$$
(7.15)

Equation (7.16) gives the error in the tie-line power flow from area 1 to area 2, which is the difference between the actual and scheduled value of the tie-line power:

$$\Delta P_{\text{tie1,2error}} = \Delta P_{\text{tie1,2actual}} - \Delta P_{\text{tie1,2schedule}}.$$
(7.16)

Equation (7.17) gives the error in tie-line power flow from area 2 to area 1:

$$\Delta P_{\text{tie,2,1error}} = a_{12} \Delta P_{\text{tie1,2error}}$$
(7.17)

since both the areas are assumed to be identical $a_{12} = -1$.

Equations (7.18) and (7.19) give the ACE of area 1 and area 2

$$ACE_1 = B_1 \Delta f_1 + \Delta P_{\text{tie12error}}$$
(7.18)

$$ACE_2 = B_2 \Delta f_2 + a_{12} \Delta P_{\text{tie12error}}.$$
(7.19)

As there is more than one GENCO in each area, ACE signals have to be given in proportion to their participation in LFC. The coefficient that distributes the ACE to all GENCOs is called "ACE participation factor" (apf). The summation of Apf should be unity for each area. Hence, ACE participation factors for area 1 are apf_{11} and apf_{12}, and apf_{21} and apf_{22} for area 2. The generated power (or) contracted power supplied by the GENCOs in (MW) is given as shown in the following equation:

$$\Delta P_{gei} = \sum_{i=1}^{4} \sum_{j=1}^{4} \text{gpf}_{ij} \Delta P_{Loj} - \text{apf}_{ij} \Delta P_{UCi}.$$
(7.20)

7.3 Optimization Techniques and Proposed Technique

The different techniques are proposed for tuning the PI controller.

7.3.1 Controllers

The tuning of controllers is more important for any system design; if tuning is not done properly, the characteristics get affected and the system becomes unstable. For the system under consideration, a Ziegler-Nichols-tuned conventional PI controller is used.

7.3.2 PI Controller

When the change in error is high, a proportional controller will be beneficial since it increases transient performance. When the error is small, the integral control mode is effective, and the steady-state is improved. Although it has the advantage of reducing overshoot, the derivative control mode increases noise and makes the system less stable due to its high sensitivity [17]. As the load changes, the derivative mode leads the system to become unstable. Figure 7.5 shows the structure of a PI controller, with K_P and K_I denoting proportional and integral gain values, respectively [18].

In this chapter, because of simplicity, flexibility, and easy design PI controller has been chosen. Eqn (7.21) shows the mathematical model of the PI controller

$$U_{\text{PI}} = K_P \text{ACE}_i + K_I \int_0^t \text{ACE}_i dt \qquad (7.21)$$

where U_{PI} is the controlled output of the PI controller, K_P is the proportional gain, and K_I is integral gain, and ACE of the concerned area is ACE.

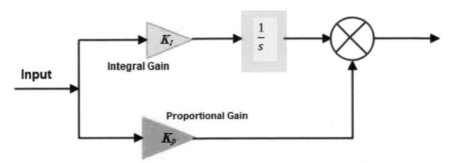

Figure 7.5   Transfer function model of PI controller.

7.3.3 Artificial Optimization Algorithm for Tuning PI

While designing the optimized PI controller, by ensuring system stability, the objective function is defined based on required specifications and parameter values [19]. In this chapter, the integral square error (ISE) is utilized as the objective function for optimizing the PI controller parameters. By employing this index, errors are decreased while the response time is increased.

The performance index for minimization of the error is shown in the following equation:

$$\text{objfn} = \text{ISE} = \int_0^{t_{\text{sim}}} (|\Delta f_1| + |\Delta f_2| + |\Delta P_{\text{tie1,2}}|)^2 dt \qquad (7.22)$$

where Δf_1 and Δf_2 are the changes in frequency of area 1 and area 2, respectively, and $\Delta P_{tie1,2}$ is the change in tie-line power flow.

7.3.3.1 Differential evolution

Differential evolution (DE) is capable of solving real-world problems that are nonlinear, noncontinuous, and nondifferential. DE places a greater emphasis on mutation than recombination or crossover when compared to other population-based meta-heuristic techniques [20]. It mutates vectors by selecting a pair of vectors from the same population at random. The mutation directs the vector toward the global optimum. The difference between two randomly picked vectors has a distribution determined by these vectors' distribution. As a result, DE can work more reliably and as a generic global optimizer [21]. DE is based on the idea of a population of vectors, with each vector representing a set of decision variables. Choice variables are chosen based on their impact on the problem to be solved. These choice variables must be encoded, and a set of starting values must be picked from the solution space. Mutation and recombination are used to create new vectors. The selection technique determines the optimal vectors based on the selection criterion [22–24].

A. Encoding

Encoding is defined as the process of converting a set of decision variables into a vector and an objective function into a fitness function. The capacity of DE to cope with floating-point and mixed-integer data simplifies the process of encoding decision variables into vectors. The number of option variables determines the size of the vector, and each vector represents a single solution from the problem's solution space.

B. Mutation

The goal of mutation is to increase search variety in the parameter space while also directing current vectors with a reasonable level of parameter variation in a way that produces better results at the right moment. It maintains the search's robustness while also exploring new parts of the search area. There are four types of mutation [8]. DE/rand/1/bin – $Y_i = X_{r1} + F^*(X_{r2} - X_{r3})$

DE/rand/2/bin – $Y_i = X_{r1} + F^*(X_{r2} - X_{r3}) + F^*(X_{r4} - X_{r5})$

DE/best/1/bin – $Y_i = X_{best} + F^*(X_{r1} - X_{r2})$

DE/best/2/bin – $Y_i = X_{best} + F^*(X_{r1} - X_{r2}) + F^*(X_{r3} - X_{r4})$

$r1 \neq r2 \neq r3 \neq r4 \neq r5$ are randomly selected.

C. Recombination

By creating offspring individuals from current individuals or vector parameters, recombination or crossover tries to reinforce previous successes. If the freshly formed individual is to be recombined, the cross-over constant is employed. Binomial and exponential cross-over are the two forms of crossover. In the binomial approach, a random number is created to form the trailing vector. The adjusted vector variable is used if the value is smaller than the cross-over constant; otherwise, the target vector variable is used.

D. Selection

The fitness of the trail and target vectors is compared, and the vector with the lowest objective value for the next generation is chosen. This assures that the population size does not fluctuate over time.

7.3.3.2 Flower pollination algorithm

Flower pollination algorithm (FPA) is a population-based algorithm inspired by nature. The fundamental goal of this FPA is to ensure that plant species reproduce optimally by surviving the fittest of flowering plants [25, 26]. There are millions of flowering plants in our universe, with flowering species accounting for 80% of the total. A flower's ultimate goal is to proliferate by pollinating other flowers. Pollen transfers from one flower to another on the same plant (self-pollination-abiotic) or another plant (cross-pollination-biotic) [27]. This transformation occurs by pollinators such as wind, birds, insects, bats, and other animals. FPA performs better when compared with others in terms of accuracy and convergence speed.

The following four rules were employed to explain the concept of flower pollination.

1. Cross and biotic pollination are considered global pollination, and pollinator movement is compared to levy flight movement.
2. Abiotic and self-pollination take place as local pollination.
3. Pollinators such as birds and insects acquire flower constancy, which is proportional to the resemblance of the two flowers involved and equal to the reproduction likelihood.
4. The chance of switching from local to global pollination or vice versa can be regulated, $p = 0.7$.

In global pollination, flower pollen is carried by pollinators like birds, wind, and insects. This global pollination, i.e., rules 1 and 3 can be written as in the following equation:

$$x_i^{k+1} = x_i^k + \gamma L\left(\lambda\right)\left(g^* - x_i^k\right) \tag{7.23}$$

where x_i^k is the flower i at iteration k, and g^* is the current best solution among the solutions for the current iteration. Here, γ is the scaling factor used to control the step size and its value is $0.3\ L\left(\lambda\right)$. Pollination strength is measured by the step size parameter in specific levy-flights movements. Levy distribution is complicated by the fact that pollinators fly over a great distance with varying distances. It is used as shown in the following equation:

$$L \approx \frac{\lambda\Gamma\left(\lambda\right)\sin\left(\pi\lambda/2\right)}{\pi}\frac{1}{S^{1+\lambda}}, \left(S \gg 0\right) \tag{7.24}$$

where $\Gamma(\lambda)$ is the standard gamma function and levy distribution will be valid for longer steps $S > 0$.

Rules 2 and 3 are for local pollination and they are shown in the following equation:

$$x_i^{k+1} = x_i^k + \varepsilon\left(x_j^k + x_m^k\right) \tag{7.25}$$

where ε is local random whose values lie between 0 and 1.

7.3.3.3 Hybrid algorithm

A superior controller is required to control the frequency and tie-line power deviations; hence, to improve the system dynamic performance, a novel optimization technique is needed [28].

A. Requirement for Hybrid Algorithm

A hybrid algorithm combines two or more approaches to solve the problem, picking one (depending on the data) or switching between them throughout the procedure.

B. Hybrid DE-FPA

DE focuses on vector mutation by using a pair of vectors from the same population that is randomly selected. The search for the global best solution becomes more diverse as the number of mutations increases.

FPA is a simple and efficient pollination procedure with no mutation. DE mutation is merged with FPA to get a hybrid DE-FPA method to solve the LFC in the deregulated environment.

7.3.3.4 Design of a hybrid DE-FPA algorithm for LFC

DE-FPA is implemented in deregulated as follows.

Step 1: Initialize the flower population (variables of LFC).

Step 2: Find each flower's fitness function and get the global flower that has minimum ISE from the population.

Step 3: For each flower, initiate a random number.

Step 4: If the random number is below switch probability, go to step 5; else, apply cross-pollination (global) and go to step 6.

Step 5: Implement self-pollination (local).

Step 6: Execute mutation for enhancement of diversity.

Step 7: Repeat steps 2-6 till the stopping criterion is satisfied.

Step 8: Get the fittest solution.

7.4 Simulation Results and Discussion

Case Study – Bilateral Contract Between Areas:

Here, a DISCO may transact power from a GENCO of its area or with other areas. Equation (7.26) shows DPM which is considered participation of GENCOs in LFC is shown by the following ACE participation factor Apfs in two areas, i.e., $Apf_{11} = 0.75$, $Apf_{12} = 1 - 0.75$, $Apf_{21} = 0.5$, $Apf_{22} = 1 - 0.5$.

DISCO's demands (pu MW) are $\Delta P_{Lo1} = \Delta P_{Lo2} = \Delta P_{Lo3} = \Delta P_{Lo4} = 0.1$.

$\Delta P_{UC1} = 0$ and $\Delta P_{UC2} = 0$. In this case, uncontracted load is zero. Hence, there is no contract violation. $\Delta P_{tie1,2schedule} = -0.02$ (pu MW), using Equation (7.26) scheduled tie-line error under steady-state.

$$DPM = \begin{bmatrix} 0.5 & 0.25 & 0.5 & 0.3 \\ 0.2 & 0.25 & 0.2 & 0 \\ 0 & 0.25 & 0.2 & 0.7 \\ 0.3 & 0.25 & 0.1 & 0 \end{bmatrix}. \tag{7.26}$$

GENCO's power generation (pu MW) by using Equation (7.3) is shown below:

$\Delta P_{ge1} = 0.1545$, $\Delta P_{ge2} = 0.0665$, $\Delta P_{ge3} = 0.135$, and $\Delta P_{ge4} = 0.0655$.

Figures 7.6 and 7.7 show the frequency deviation (Hz) in areas 1 and 2. Table 7.1 gives a comparison of PI and hybrid DE-FPA tuned PI controllers in terms of undershooting, overshoot, and settling time. From the dynamic responses and comparison table, conventional PI controllers have more oscillations and overshoot. FPA-PI is superior and faster than PI controllers. Adding DFIG in both areas stabilizes the frequency response of the system.

From Figure 7.6, it is seen that the settling time for PI controller is 47 seconds, whereas for DE-FPA tuned PI controller, it is 11 seconds. Similarly, from Figure 7.7, it is observed that the settling time of the DE-FPA tuned PI controller is 9 seconds, whereas for the conventional PI controller, it is 35 seconds. As a result, the hybrid DE-FPA tuned PI controller is better when compared with the PI controller.

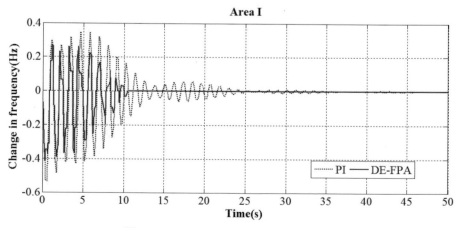

Figure 7.6 Frequency deviations in area 1.

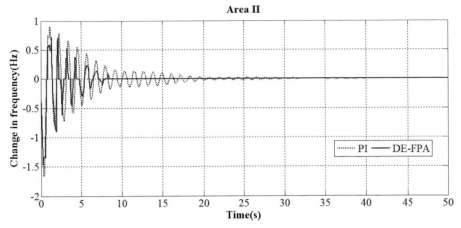

Figure 7.7   Frequency deviations in area 2.

Table 7.1   Frequency deviation of each area.

Controllers used	Area	Peak shoot (Hz)	Undershoot (Hz)	Settling time (seconds)
		Frequency deviation (Hz)		
PI	1	0.38	−0.52	47
	2	0.98	−1.7	43
DE-FPA	1	0.28	−0.41	11
	2	0.7	−1.45	9

From Table 7.1, the performance of the PI and DE-FPA tuned PI controllers in area 1 declares that with peak overshoot and peak undershoot as the performance index; the DE-FPA tuned PI controller has 27% and 26% improvement over the conventional PI controller. The settling time of the DE-FPA tuned PI controller is around 82% less than the PI controller.

Similarly, for area 2 with peak shoot and undershoot as the performance index, the DE-FPA tuned PI controller showed an improvement of 27% and 23% over conventional PI controllers. The settling time of the DE-FPA tuned PI controller is found to be 77% better when compared with the PI controller.

7.5 Conclusion

The deregulated power system has bilateral transactions for the effective utilization of available resources. The RESs are intermittent for the continuity of the power supply. Wind energy systems for electric power generation are considered in this work. The major issue to solve in the bilateral transaction violation during practical aspects is that it depends on the demand variation and available resources. The solution to identify and solve this is the measurement of frequency deviation in the areas where the violation is taking place. For the deregulated power system two-area LFC with wind energy generator having a DFIG simulation model is developed. The frequency deviation of areas 1 and 2 is measured and controlled by using a PI controller. Conventional techniques tuning to find the coefficient of PI controller are measured. Hybrid intelligent algorithm DE-FPA is developed and used to find the coefficient of PI controller for the better performance of the LFC. The performance measures are overshoots, settling time, and oscillations before settling. Developed hybrid intelligent algorithms give better performance compared with the conventional PI controller. The frequency deviation waveforms of area 1, area 2, and tie lines are discussed for the conventional and hybrid intelligent algorithm. The hybrid intelligent algorithm DE-FPA gives better performance.

Appendix

Table 7.2   System parameters.

Parameters, symbols (units)	Values
Governor time constants $G_j(s)$	0.08
Rating, P_{ri} (MW)	2000
D_i system damping coefficient (pu MW/Hz)	8.333×10^{-3}
Power system time constant, T_{Pi} (s)	20
Turbine time constants, T_{Tj} (s)	0.3
Gain constant of power system, K_{Pi} (Hz/pu MW)	120
Bias factor, B_i (pu MW/Hz)	0.425
Speed regulation, R_j (Hz/pu MW)	2.4
Inertia constant, H_i (s)	5
Tie-line power constant, T_{12}	0.0826
DFIG gain constant (K_{wi1}, K_{wi2})	0.5
DFIG gain constant (K_{wp1}, K_{wp2})	1
System frequency, f (Hz)	60
Wind inertia constant (H_{e1}, H_{e2}) (pu MW s)	3.5
DFIG time constant (T_{a1}, T_{a2}) s	0.2
DFIG time constant (T_{r1}, T_{r2}) s	15
DFIG time constant ((T_{w1}, T_{w2}) s	6

References

[1] lgerd, O. I. "Electric energy systems theory: An introduction." 1982.

[2] Kumar, J., Ng, K-H., and Sheble, G. "AGC simulator for price-based operation-part II: Case study results." *IEEE Transactions on Power Systems* 12.2 (1997): 533–538.

[3] Christie, R. D., and Bose, A. "Load frequency control issues in power system operations after deregulation." *IEEE Transactions on Power Systems* 11.3 (1996): 1191–1200.

[4] Donde, V., Pai, M. A., and Hiskens, I. A. "Simulation and optimization in an AGC system after deregulation." *IEEE Transactions on Power Systems* 16.3 (2001): 481–489.

[5] Aditya, S. K., and Das, D. "Battery energy storage for load frequency control of an interconnected power system." *Electric Power Systems Research* 58.3 (2001): 179–185.

[6] Selvaraju, R. K., and Somaskandan, G. "Impact of energy storage units on load frequency control of deregulated power systems." *Energy* 97 (2016): 214–228.

[7] Sasaki, T., Kadoya, T., and Enomoto, K. "Study on load frequency control using redox flow batteries." *IEEE Transactions on Power Systems* 19.1 (2004): 660–667.

[8] Chidambaram, I. A., and Paramasivam, B. "Optimized load-frequency simulation in restructured power system with redox flow batteries and interline power flow controller." *International Journal of Electrical Power & Energy Systems* 50 (2013): 9–24.

[9] Shayeghi, H. A. S. H., Shayanfar, H. A., and Jalili, A. "Load frequency control strategies: A state-of-the-art survey for the researcher." *Energy Conversion and Management* 50.2 (2009): 344–353.

[10] Tyagi, B., and Srivastava, S. C. "A LQG based load frequency controller in a competitive electricity environment." *International Journal of Emerging Electric Power Systems* 2.2 (2005).

[11] Zamani, A. A., *et al.* "Optimal fuzzy load frequency controller with simultaneous auto-tuned membership functions and fuzzy control rules." *Turkish Journal of Electrical Engineering & Computer Sciences* 22.1 (2014): 66–86.

[12] Zamani, A. A., *et al.* "Optimal fuzzy load frequency controller with simultaneous auto-tuned membership functions and fuzzy control rules." *Turkish Journal of Electrical Engineering & Computer Sciences* 22.1 (2014): 66–86.

[13] Demiroren, A., and H. L. Zeynelgil. "GA application to optimization of AGC in three-area power system after deregulation." *International Journal of Electrical Power & Energy Systems* 29.3 (2007): 230–240.

[14] Bhatt, P., Roy, R., and Ghoshal, S. P. "Optimized multi area AGC simulation in restructured power systems." *International Journal of Electrical Power & Energy Systems* 32.4 (2010): 311–322.

[15] Fathima, A. P., and Abdullah Khan, M. "Design of a new market structure and robust controller for the frequency regulation service in the deregulated power system." *Electric Power Components and Systems* 36.8 (2008): 864–883.

[16] Zahira, R., *et al.* "Modeling and simulation analysis of shunt active filter for harmonic mitigation in islanded microgrid." *Advances in Smart Grid Technology.* Springer, Singapore, 2021. 189–206.

[17] Lakshmi, D., and Zahira, R., "Load frequency control in deregulated power system." *International Journal of Research in Arts and Science* 5 (2019): 124–133. Special Issue in Holistic Research Perspectives.

[18] Lakshmi, D., *et al.* "Flower pollination algorithm in DPS integrated DFIG for controlling load frequency." *2020 International Conference on Power, Energy, Control and Transmission Systems (ICPECTS).* IEEE, 2020.

[19] Yang, X.-S. *Engineering Optimization: An Introduction with Meta-heuristic Applications.* John Wiley & Sons, 2010.

[20] Jin, L., *et al.* "Robust delay-dependent load frequency control of wind power system based on a novel reconstructed model." *IEEE Transactions on Cybernetics* (2021).

[21] Panda, A., et al. "A PSO based PIDF controller for multiarea multisource system incorporating dish stirling solar system." *2019 International Conference on Intelligent Computing and Control Systems (ICCS).* IEEE, 2019.

[22] Tang, J., Liu, G., and Pan, Q. "A review on representative swarm intelligence algorithms for solving optimization problems: Applications and trends." *IEEE/CAA Journal of Automatica Sinica* 8.10 (2021): 1627–1643.

[23] Zahira, R., and Lakshmi, D. "Control techniques for improving quality of power-A review." *International Journal of Research in Arts and Science* 5 (2019): 107-123. Special Issue in Holistic Research Perspectives [Volume 4].

[24] Pahadasingh, S. "TLBO based CC-PID-TID controller for load frequency control of multi area power system." *2021 1st Odisha International Conference on Electrical Power Engineering, Communication and Computing Technology (ODICON)*. IEEE, 2021.

[25] Chen, X., *et al.* "A survey of swarm intelligence techniques in VLSI routing problems." *IEEE Access* 8 (2020): 26266–26292.

[26] Zahira, R., Lakshmi, D., and Ravi, C. N. "Power quality issues in microgrid and its solutions." *Microgrid Technologies* (2021): 255–286.

[27] Rehman, U., *et al.* "Load frequency management for a two-area system (thermal-PV & Hydel-PV) by swarm optimization based intelligent algorithms." *2021 International Conference on Emerging Power Technologies (ICEPT)*. IEEE, 2021.

8

Application of Machine Learning Techniques in Modern Hybrid Power Systems – A Case Study

B. Koti Reddy[1], Krishna Sandeep Ayyagari[2], Raveendra Reddy Medam[3], and Mohemmed Alhaider[4]

[1]Department of Atomic Energy, India
[2]Department of Electrical & Computer Engineering, The University of Texas at San Antonio, USA
[3]Department of EEE, Maturi Venkata Subba Rao Engg (MVSR) College, India
[4]College of Engineering at Wadi Addawaser, Prince Sattam bin Abdulaziz University, Saudi Arabia
E-mail: kotireddyb@ieee.org; krishnasandeep.ayyagari@my.utsa.edu; raveendra_eee@mvsrec.edu.in; malhider3@gmail.com

Abstract

Due to the rapid adoption of intelligent power electronic devices and digital technologies, traditional vertically designed power systems are being phased out and replaced by modern hybrid power systems. Since its inception, the power system has undergone numerous changes that have increased system efficiency, increased the share of renewable energy, and made it easier to control. However, such a rapid revolution in electrical power systems during the current Industrial Revolution has increased its complexity. The primary concerns are cybersecurity, forecasting supply and demand, optimal power allocation, power quality maintenance, and a skilled workforce shortage. Digital tools aid in load management and the optimization of various power resources. Modern hybrid power systems, artificial intelligence techniques such as machine learning and optimization algorithms, are emerging in the power sector for better control. Nonetheless, little research has been conducted on machine learning applications in industries with integrated power resources. Machine learning techniques will be used in the industries to forecast supply and demand, make the best use of energy resources, etc. This chapter aims to discuss the use of machine learning techniques in modern hybrid power systems. A well-known large industry with multiple energy resources has been considered for this purpose. All components of the power system network are modeled, and simulations are run to determine the best way to use them under various generation and load scenarios, weather conditions, and financial conditions. For the case study considered, the simulated results are validated using field data and ETAP software, and the results are encouraging.

Keywords: Energy resources, machine learning (ML), artificial intelligence (AI), optimization, ETAP, modern hybrid power systems (MHPS).

Notation: Upper-case (lower-case) boldface will denote matrices (column vectors); $(\cdot)^T$ for transposition; $(\cdot)^*$ for complex-conjugate; and $(\cdot)^{-1}$ for inverse. Re denotes the real part of the complex number, and $j := \sqrt{-1}$ the imaginary unit. For a given $N \times 1$ vector x, $\mathrm{diag}(x)$ returns $N \times N$ matrix with the elements of x in its diagonal. Finally, I_N denotes $N \times N$ identity matrix; and $0_N, 1_N$ the N-dimensional vectors with all zeroes and ones, respectively, and $0_{N \times M}$ is $N \times M$ matrix with all zeroes.

8.1 Introduction

The electrical power system (EPS) has undergone various modifications since its inception. This is due to rapid technological improvements, computer percolation, and intelligent electronic devices (IED) in EPS. Furthermore, while the rising use of renewable energy (RE) such as wind and solar may have given conventional EPS a facelift, it has made the system more complex due to its intermittency and reliance on weather conditions [1]. Figure 8.1 depicts the main distinction between standard EPS and modern hybrid power system (MHPS).

In addition, the EPS has modified its previous vertical shape of power flow from top (generation) to bottom (consumption) (T2B) with bottom-to-top (B2T) approach of inventory flow from consumer to generator to a two-way approach, as illustrated in Figure 8.2.

Furthermore, the evolution of de-regulated electricity markets (DEM), distributed generation (DG), microgrids (MGs), smart grids (SGs),

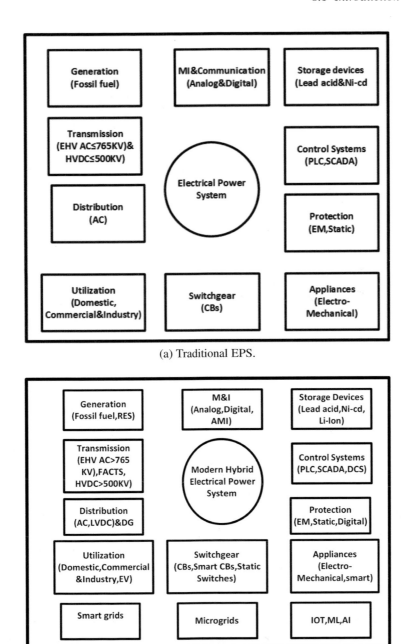

(a) Traditional EPS.

(b) MHPS.

Figure 8.1 Electrical power system. (a) EPS. (b) MHPS.

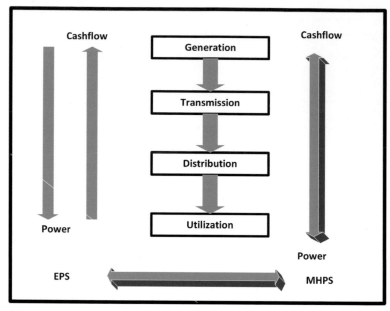

Figure 8.2 Power and inventory flow (traditional at left and MHPS at right).

smart appliances (SAs), and energy storage systems (ESS) has changed the name of EPS to MHPS. Moreover, there are many expectations on EPS globally in terms of the five Es, which are energy, efficiency, environment, equipment, and economics. In addition to meeting these expectations, the MHPS provides many benefits to humanity, including the utilization of abundantly available renewables, de-centralized EPS, reduced operation and maintenance costs, reduced losses, environmental protection, and so on. However, such a rapid revolution in EPS during the current Industrial Revolution (IR 4.0) has increased its complexity. The technical issues involved in MHPS, as well as their mitigation strategies, are discussed in the following sections of this chapter.

8.2 Technical Issues in Modern Hybrid Power Systems

The accelerated advancement of EPS has resulted in rapid industrial development and the extension of power supply to every corner of the globe, with an increased share of renewables. However, due to the three Ds (de-carbonization, de-centralization, and digitalization), this type of paradigm shift has resulted in the following technical issues.

8.2.1 Power Quality

The power quality (PQ) issues related to the MHPS are as follows:

1. voltage variations and unbalance among phases in case of poly-phase systems;
2. low voltage and high voltage ride through [2];
3. frequency fluctuations;
4. harmonics due to the presence of power electronic devices (PEDs) [3];
5. voltage transients;
6. voltage flickering, swell, and sags;
7. reactive power management especially in case of inverter-operated power systems having no-reactive power support [4].

8.2.2 Demand–Supply Management

Here, the main issues are as follows:

1. load forecasting;
2. weather forecasting;
3. load shedding;
4. balance between demand and supply;
5. optimal allocation, and operation of various energy resources and storage systems [5].

8.2.3 Synchronization and Islanding

Here, the major issues to be dealt with are as follows:

1. connection and re-connection of various resources with proper synchronization;
2. anti-islanding facility and islanding operations;
3. maintaining stability of generators.

8.2.4 Protective Devices, Safety, and Environment

The important safety-related issues are as follows:

1. operational strategies;
2. maintenance of equipment;
3. correct operation of protective devices;
4. proper coordination among various protective devices;
5. environmental issues with respect to use of semiconductors, and energy storage batteries;

6. estimation of proper settings of protective devices [6];
7. perfect grounding system;
8. reverse power flow conditions;
9. island and grid-interactive operational procedures;
10. maintenance of reliability due to lack of proper sizing and mis-positioning of DGs.

8.2.5 Human Factor

The main issues are as follows:

1. availability of skilled manpower;
2. training and retaining of skilled manpower;
3. re-training and skill development of human resources.

8.3 Application of ML and Optimization Techniques in MHPS

Most of the technical issues mentioned above can be mitigated with the help of machine learning (ML). Supervised learning (classification and regression) and unsupervised learning (clustering) are the two main types of ML [7]. Regression, support vector machines (SVMs), artificial neural network (ANN), and other common algorithms all use supervised learning. Unsupervised learning employs K-means clustering, predictive maintenance, component analysis, and other techniques. The necessary steps for the successful development of an ML model are described in the box given below.

Steps in the development of a machine learning method:

Collect data for the purpose of prediction

Data should be pre-processed and in the proper format.

Investigate data- to learn about insignificant values, errors, and so on.

Divide the data into training (75 %) and testing (25 %)

Train the algorithm with training data until it produces a correct method with the fewest errors.

With testing data, put the model to the test. Examine the outcomes

Table 8.1 ML application in MHPS.

Sl. no.	MHPS component	Application examples of ML
1	Generation	• Renewable energy integration [8] • Optimal allocation of power [9] • Forecasting of energy from weather-dependent resources such as solar and wind [10]
2	Transmission	• Optimal allocation of transmission lines and its assets • Implementation of FACTS (flexible AC transmission system) [11] • Use of UHV and EHV DC systems • Monitoring of lines at remote or hill station or higher elevations
3	Distribution	• Integration of distribution generation • Optimal operation of microgrids and smart grids • Adaptive protection system of DG, MG, and SGs [12] • Integration of smart appliances with system • Optimal allocation of energy resources • Effective load shedding as and when required [13] • Occupancy sensors and automation of HVAC functions
4	Others	• Augmentation and optimum utilization of other assets • Safety and security of human and assets • Demand-supply forecasting and its management • Predictive maintenance of equipment [14] • Energy storage system management [15] • Optimal cost of operation and maintenance • State evaluation and fault detection [16]

The paradigm shift in EPS needs new technologies such as ML and artificial intelligence (AI), whereas ML is the way to achieve AI. ML is used in various constituents of MHPS as shown in Table 8.1.

8.4 A Prediction Case Study of ML in MHPS

Digital tools aid in the better management and optimization of various energy resources, particularly RERs. AI techniques like ML are emerging in the power sector to improve RE prediction, which is mostly intermittent. To rely on RERs for optimization and better control, industries must be able to predict an hour or a day ahead of time power with varying weather conditions. For this purpose, a case study of a well-known large-scale continuous process chemical industry that has both a cogeneration power plant (CPP) and a solar photovoltaic power plant (SPP) as DG is used to forecast solar irradiance using the ML technique. Figure 8.3 depicts a single line diagram (SLD) for the case study industry. The following are the details for the MHPS.

This section describes the power distribution network, CPP setup, and the SPP details for the case study considered.

A coal-fired steam boiler is installed to meet the process plant's need for high-pressure steam at higher temperatures. Excess or waste steam is used to power a turbine-generator (TG) set. The process plant's total power

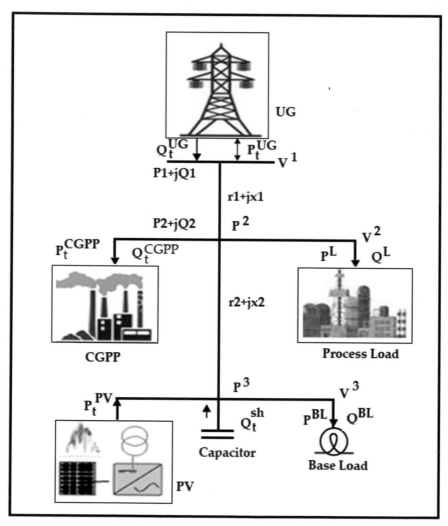

Figure 8.3   Single line diagram of the industry.

requirement is 18 MW, with a process load demand of 15 MW required during normal industry operation. The remaining 10 MW is always required to support plant base loads like water systems and ventilation. When the CPP is turned off, the utility grid will supply the base loads of 10 MW using a demand management controller, which is used to control the unwanted demand spikes caused by the large induction motors' sudden turn-on and turn-off. As a continuous process chemical industry, it runs almost continuously if the industry is operational. Furthermore, during the night, some of the loads that run during the day are turned off. During the night, however, the lighting load will balance the net load. As a result, the load profile over the course of a full day is constant. A 2.3-MVAr shunt capacitor is connected to a 6.6-kV bus to cater plants' reactive power load.

CPP with Operational Issues:

In the mentioned industry, a CPP with a steam boiler and a TG is installed. A radiant boiler with integral superheaters, a forced flow section, and a tubular air heater with two forced draft and two induced draft fans comprise the steam generator. The turbines are extraction condensing and have TG. The TG is a two-pole, three-phase, air-cooled motor that is directly connected to the steam turbine. Their maximum output is rated at 6.6 kV and is rated at 18 MW at 0.85 power factor. In addition, as shown in Figure 8.3, the CPP is linked to a utility grid for initial operation. It is worth noting that the minimum power output of each turbine, from an operational standpoint, would be at least 60% of the 18 MW (i.e., 10.8 MW). The TG will provide 18 MW of active power, which is the industry's exact requirement. Later, the industry installed a 6-MW peak capacity SPP to meet statutory requirements and to promote RE. Furthermore, during the SPP generation time (approximately 5-6 hours per day), the CPP operation is kept running at a lower output of around (13 MW), with SPP supplying the remainder to satisfy power balance. As a result of the reduction in CPP output, the *overall efficiency of the steam turbine falls* from 34% to 32%, and even the extraction of steam into the process system becomes more complicated.

Solar Photovoltaic Power Plant with Operational Issues:

The total SPP was divided into two divisions of 3 MW each. A total of 23,530 PV modules with a peak power rating of 255 W, open circuit, and nominal voltages of 37.6 and 30.5 V, respectively, short circuit and nominal currents of 8.95 and 8.42 A, respectively, panel efficiency of 15.5%, and annual power de-rating of 4% [17] were installed with six 1-MW inverters. These inverters are grid-interactive and do not have reactive power capability. The power was evacuated at the 6.6-kV level. The levelized cost of energy was estimated to be US $ 0.1 (in July 2021).

The following operational interlocks are required for the joint management of the CPP and SPP. If the CPP's TG is tripped, the industry continues to operate with assistance from the utility grid and SPP when it is available. If the process load is suddenly reduced due to any abnormality in process operations, the CPP will be islanded, and the SPP will be kept *off* to avoid unauthorized power export to the grid and to prevent sudden voltage rise at plant buses. Circuit breakers have an additional mechanical interlock for this purpose (CB). When the CPP is not producing power, the SPP trips within 0.08 seconds (i.e., the trip time of CBs at CPP and SPP together). This is well within the 0.16 seconds specified by [18]. Even though such local PV integration reduces power losses by 3%-9% [19], this scenario indicates a certain curtailment of RE; in this case, PV, for the efficient and secure operation of power systems that are integrated with grids [20]. The design efficiency of a CPP is 34% up to 16 MW, and it drops to 32% when the load on the CPP is 14 MW. This is primarily due to the operation of SPP. With the operational issues of CPP and SPP, there is an urgent need to find an optimal solution for such industries. It is also necessary to investigate irradiance forecasting techniques for a large-scale PV plant. This is the theme of the ensuing subsection.

8.4.1 Forecasting Irradiance of SPP

The schematic diagram of 6-MW SPP is shown in Figure 8.4 The optimization of EPSs with DG is typically done with prior information on load profile and weather forecasting data to ensure satisfactory operation. The prediction of solar power generation for a day or a specific time ahead based on such weather forecasts will undoubtedly aid power generation stations in planning and dispatch. The prediction should be automatic and more accurate by considering seasonally varying weather conditions, which is possible with the help of ML techniques, as used by many researchers [21, 22]. However, every model has limitations, and there are some deviations. Figure 8.5 depicts a dataset of every hourly reading on March 1, 2021, at the case study industry's site location (latitude of 17.90 N and longitude of 80.80 E) which amounts to 300 datapoints. The ambient temperature and relative humidity are input variables in this data, and the output variable is global horizontal

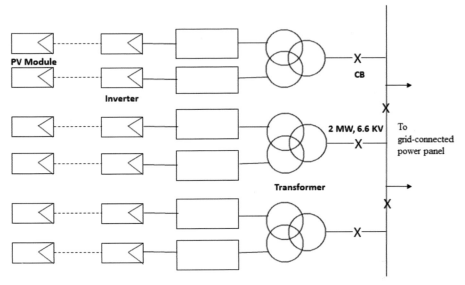

Figure 8.4 Schematic diagram of 6-MW SPP.

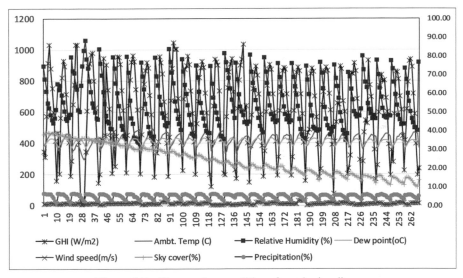

Figure 8.5 Site weather conditions for solar irradiance.

irradiance (GHI). As pointed out earlier, the performance of predictor plays an important role in the operation of the CPP. The useful measures necessary to evaluate the predictions' performance are listed in the ensuing subsection.

8.4.2 Metrics for Understanding the Performance of Predictions using ML Methods

Some of the metrics used in regression analysis are briefly explained here [23].

1. *Mean absolute error* (MAE) is the absolute difference between the true values and the predicted values. Since it is absolute, any negative sign in the result is to be ignored. It takes the average of error from each sample, as in the following equation:

   MAE= True values – Predicted values

$$= \sum |y - \hat{y}|/n \tag{8.1}$$

   where y = true value
   y = predicted value and
   n = number of observations.
   It is generally used for continuous variable data and the lower value of MAE gives a better regression model. It treats the errors equally whether it is large or small.

2. *Mean squared error* (MSE) is estimated by considering the averages of the squares of the difference between the true values and the predicted values of the dataset as in the following equation:

$$\text{MSE} = 1/n \sum (y - \hat{y})^2. \tag{8.2}$$

   Here, by squaring the difference between the true values and the predicted values, the higher error value can be penalized. It is the most widely used metrics when the dataset contains too high or too low values, but it is not useful when the data contain huge noise. It is the variance of the error value.

3. *Root mean squared error* (RMSE) is the square root of MSE, which gives better accuracy of the regression model, as in the following equation:

$$\text{RMSE} = \sqrt{\frac{\sum (y - \hat{y})^2}{n}}. \tag{8.3}$$

   It is more useful when large errors are present in the dataset. It is also the standard deviation of errors.

4. *R-squared*: It is the coefficient of determination and indicates the qualitative information of a dataset, as in the following equation:

$$R^2 = \frac{\text{SSEM} - \text{SSER}}{\text{SSEM}}. \tag{8.4}$$

Here, SSEM is the sum of squared errors by mean line and SSER is the sum of the squared errors by regression line. Here, the amount of error is reduced since the regression model is used instead of the mean of mean line. Its value lies between 0 and 1 where the value 1 or close to 1 gives a better performance regression model, i.e., a perfect model and a value equal to zero indicates that the model does not fit for the given data.

5. *P-value*: It indicates the statistical significance of coefficient relationship of a model. A lower value of less than 0.05 indicates a meaningful addition to the dataset. It tests the *null hypothesis* of each term of coefficients. A *lower* p-value means a significant change in prediction value for a change in response variable. It gives a direction of which term to be retained in a regression model.

8.4.3 Model-Based and Model-Free Regression Techniques

In general regression analysis (supervised learning), the statistical relationship between a dependent variable and one or more independent variable are estimated. Given k training points are $(x_1, y_1), \ldots, (x_k, y_k)$.

The *model-based* regression analysis solves the following optimization problem, as in the following equation:

$$(\text{P1}) \quad \min_{\{\beta\}} \frac{1}{n} \sum_{\{i=1\}}^{\{k\}} \left(\beta^T x_i - y_i\right)^2. \tag{8.5a}$$

Upon solving the optimal β, the test points are then used to obtain the estimated values $(\hat{y}_1, \ldots, \hat{y}_k)$ to unforeseen inputs $(x_1, \ldots, x_n)$. Note that the problem (P1) is differentiable with respect to β and can be solved via off-the-shelf convex optimization techniques. In this study, the predictions from regression from problem (P1) are termed as *model-based* because the coefficients of the input variables are exactly known.

The attention is then turned to the prediction techniques by ANNs which are generally termed as *model-free*.

ANNs are widely used in a wide range of engineering applications, including forecasting, modeling complex patterns, image recognition, and regression. Solar radiation prediction models are developed using ANNs in this work. A neuron is the basic unit of ANN, and it generates output using a transfer function. Each input will be multiplied by a weight, which will act as a link between the input and the neuron as well as between the various layers of neurons. The neuron uses a transfer function to obtain the result in the final stage of ANN. The advantage of ANN techniques is that they do not require knowledge of mathematical calculations between parameters *(model-free)*, but they do require less computational effort and provide a compact solution for multivariable problems.

The ANNs find a nonlinear mapping between the training and test inputs given as $\Omega : \varphi_k(x_k) \rightarrow y_k$. For simplicity, let vectors $\mathbf{d_k}$, $\theta_\mathbf{k}$ and $\mathbf{b_k}$ represent the trainable parameters of the ANN. In the task of supervised learning, the ANN is trained using back-propagation algorithm based on gradient descent which minimizes the training loss given as in the following equation:

$$(\text{P2}) \quad \min_{\theta_\mathbf{k}} \sum_k ||y_k - g_k\left(\mathbf{d_k}, \ \theta_\mathbf{k}\right)||^2 . \qquad (8.5b)$$

Upon training, the ANN finds the optimal parameters $(\theta^*, \mathbf{b}^*)$ for each layer. The trained ANN with optimal parameters is then used to estimate the values for the unanticipated inputs [24]. When compared to *model-based* techniques, ANNs *(model-free)* can model any degree of nonlinearity. However, in practice, it is vital to evaluate the performance of both *model-based* and *model-free* prediction strategies. Because the investigated case study is a continuous-running process plant, the optimal judgments made while accounting for predicting mistakes in the GHI play an important part in decision making. As a result, the prediction accuracies of both *model-based* and *model-free* models are analyzed in the next section. Figure 8.6 describes the complete methodology of the optimal decision making in the MHPS using *model-based* and *model-free* approaches.

8.4.4 Prediction Block

This block preprocesses the training dataset, which consists of one month of data with appropriate inputs required to predict GHI, and then trains the corresponding *model-based* and *model-free* regression models. The trained models are then used to estimate/predict the GHI on test day. Furthermore, it is assumed that the models are trained offline, which means that when the

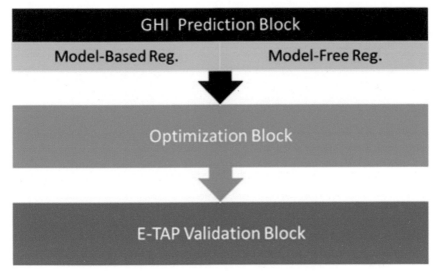

Figure 8.6 Complete methodology for optimal decision making.

site conditions change, the training must be repeated, and the corresponding models are updated using the cloud software installed in the process plant. Upon getting the prediction of GHI from *model-based* and *model-free* regression models, the best option is to be selected based on the following criteria for each method:

1. smaller value of RMSE, MSE, and MAE;
2. *R*-squared value closer to 1.0;
3. simplicity and faster response.

8.4.5 Forecasting of Solar Irradiance with a Model-Based Regression Approach

In the current study, one-month data collected in the plant location used for the *model-based* regression approach and simulations carried out on various regressions models and the results are shown in Table 8.2.

With the above criteria, the linear regression-linear with *R*-squared of 0.76, RMSE of 131.31 and MSE of 17243 with a fastest response time of 0.97 seconds is selected as the best fit prediction model. The regression results are shown in Table 8.3.

In this selected *model-based* regression model, the parameters with "*p*" value of more than 0.05, i.e., wind speed and sky cover, are ignored since

Table 8.2 Comparison of parameters of various regression models.

Model	Model Parameter	RMSE	R-squared	MSE	MAE	Training Time (s)	Remarks
Linear Regression	*Linear*	*131.31*	*0.76*	*17,243*	*95.37*	*0.97*	*Our best*
	Interactions	127.72	0.77	16,312	96.82	5.15	-
	Robust	139.24	0.73	19,387	91.24	4.84	-
	Stepwise	128.34	0.77	16,470	92.10	12.01	-
Tree	Fine	156.71	0.66	20,559	99.92	2.38	-
	Medium	144.10	0.71	20,766	93.59	2.88	-
	Coarse	162.16	0.63	26,296	117.35	2.70	-
Support vector machine (SVM)	Linear	134.63	0.75	18,124	91.19	3.35	-
	Quadratic	132.40	0.76	17,526	96.80	3.57	-
	Cubic	132.57	0.75	17,576	88.97	5.23	-
	Fine Gaussian	143.38	0.71	20,559	100.86	8.07	-
	Medium Gaussian	175.20	0.78	15,675	84.18	7.80	-
	Coarse Gaussian	132.50	0.75	17,756	87.31	5.17	-
Gaussian process regression (GPR)	Rationale quadratic	116.56	0.81	13,587	81.86	14.04	-
	Squared exponential	*115.90*	*0.81*	*13,433*	*82.10*	*12.61*	*System best*
	Matern 5/2	116.34	0.81	13,356	81.31	13.05	-
	Exponential	117.84	0.81	13,886	81.35	14.16	-
Ensembles of trees	Boosted trees	131.01	0.76	17,163	90.20	10.34	-
	Bagged trees	129.65	0.73	16,808	84.40	12.93	-

they do not have much significance on irradiance values. Training the field data without wind speed and sky cover is done and the results are shown in Table 8.4. Here, a model with the *R*-squared of 0.76, RMSE of 132.06, and MSE of 17439 with a fastest response time of 0.82 seconds is selected as the best prediction model. The final prediction plot is as in Figure 8.7.

Table 8.3 Initial regression results for irradiance prediction.

Statistics	Estimate	SE	t-Stat.	p-Value
(Intercept)	−6102	505.57	−12.07	1.53×10^{-26}
Amb. temp (°C)	50.87	6.04	7.95	8.39×10^{-10}
Rel. humidity (%)	9.90	2.19	4.51	7.44×10^{-5}
Dew point (°C)	106.86	8.94	11.96	8.36×10^{-25}
Wind speed (m/s)	−33.56	17.49	-1.91	0.58
Sky cover (%)	−0.47	1.05	−0.46	0.64
Precipitation (%)	151.24	10.98	13.77	1.82×10^{-28}

Table 8.4 Final regression results for irradiance prediction.

Statistics	Estimate	SE	t-Stat.	p-Value
(Intercept)	−6244	510.92	−12.22	4.59×10^{-28}
Amb. temp ((°C))	51.94	6.37	8.15	1.04×10^{-14}
Rel. humidity (%)	10.89	2.27	4.79	2.68×10^{-6}
Dew point (°C)	106.72	9.06	11.79	1.69×10^{-26}
Precipitation (%)	151.01	11.16	13.52	8.97×10^{-33}

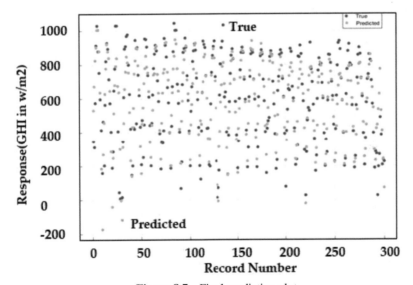

Figure 8.7 Final prediction plot.

With the above results, the relation function for solar irradiance is formulated as in the following equation:

$$\hat{G}_{\{m-\text{based}\}} = 51.94\, x_1 + 10.89 x_2 + 106.72\, x_3 + 151.01\, x_4 + (-6244)$$
$$(8.6a)$$

where x_1, x_2, x_3, and x_4 are, respectively, the ambient temperature, relative humidity, dew point, and precipitation. The intercept is given by the last term in the equation, which is -6244.

8.4.6 Forecasting of Solar Irradiance with a Model-Free Regression Approach (ANNs)

In the absence of solar radiation measuring instruments, the design of the ANN model is critical for estimating global solar radiation. This ANN model will be extremely useful in maximizing the utilization of a large amount of free, environmentally friendly solar energy for a variety of applications. The feed-forward networks are the most common type of multi-layer perceptron (MLP). A typical MLP has three distinct layers, which are referred to as the input, hidden, and output layers; in that order, $x_1, x_2,$ $x_3,$ and x_4 are the input signals. The input layer neurons, which only serve as buffers, distribute to the hidden layer neurons. Figure 8.8 [25] depicts each neuron located within the hidden layer after weighing the input signals. It adds them together with the assigned connection strengths from the layer in the input and output. The output can also be calculated using the sum function [26]. A set of input–output data can be used to train and estimate the value of an ANN model. In a multilayer feed-forward network, the most used algorithms for training are Levenberg–Marquardt (LM) back-propagation, Bayesian regularization (BR), and scaled conjugate gradient (SCG). The ANN model used in this study was built in MATLAB version 2019 and consisted of three layers of feed-forward network with *tangent sigmoid activation* function in hidden layers and *linear activation* function in output layer. Four parameters are used as inputs: ambient temperature, relative humidity, dew point, and precipitation, with only one parameter, monthly average global solar radiation (GHI), predicted as an output.

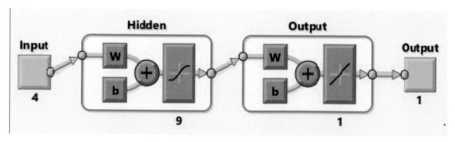

Figure 8.8   Architecture of artificial neural network [26].

8.4.7 Normalization, Training, and Testing for Model-Free Regression

The training set is used in a *model-free* regression (ANN) to learn by *optimally* adjusting the weights on the neural network as described in the preceding section. In other words, the ANN model should predict new data (unforeseen data) with as little error as possible. The testing set is only used to assess the efficacy of an ANN model that has been built. In the current study, one-month data collected in the plant location was normalized to a range between -1 and +1 before the training process [28]. After normalization, the entire dataset was divided into three categories for training (70%), testing (15%), and validation (15%) of an ANN model. Choosing the number of neurons in a hidden layer and the number of hidden layers for any ANN model is difficult. For most complex applications, one hidden layer is sufficient.

The training of an ANN model begins with a random initial weight. It is critical to identify the optimal number of neurons within the hidden layer to achieve the lowest possible error in ANN model output. Based on MSE and R^2 results, an acceptable range of neurons in the hidden layer is chosen. The optimal number of hidden neurons was determined when the MSE was the lowest and the linear correlation coefficient (R) was the highest.

To optimally find the number of hidden neurons in the hidden layer, the ANN model's training was done by increasing the number of neurons one by one until it converged into a minimum mean square error. Among the three algorithms, the R-value for the LM algorithm was in the range of 0.9000. When the value of R exceeds 0.9, it indicates that most predicted values match the measured values. As a result, the LM algorithm was considered in the current work to determine the optimal number of neurons in the hidden layer. According to the preceding discussion, nine neurons in the hidden layer were discovered to be the best at predicting GHI. Figure 8.9 depicts the regression analysis and Figure 8.10 depicts the training performance.

The GHI predicted using *model-free* regression approach can be given as in the following equation:

$$\hat{G}_{\{m-\text{free}\}} = (x_n, \ g_k(\theta^*, b^*)) \tag{8.6b}$$

It is worth mentioning that the *model-free* prediction model outperformed the *model-based* prediction model in the case study considered with a slightly higher R^2 on the training dataset. This is because the *model-free* approach

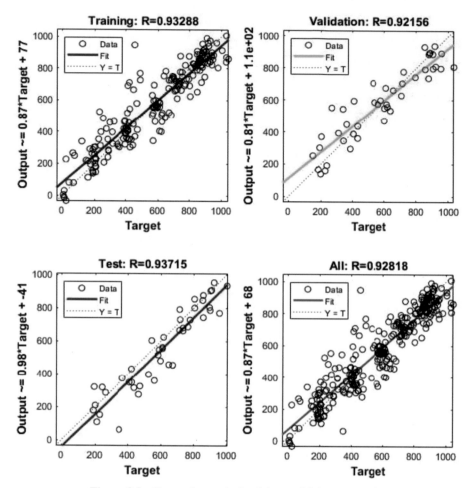

Figure 8.9 Regression analysis of the *model-free* approach.

models the nonlinear interactions between the inputs necessary to predict GHI. The key problem here is that when these models are utilized to estimate the actual test day GHI forecasts, the GHI must be estimated with as little error as possible. Because a big estimation inaccuracy can have a negative impact on the CPPs' performance. Therefore, the *optimization-assisted ML* approach is necessary to gauge the accuracies of the predictions. These predicted forecasts are now fed into the *optimization block* to determine the optimal CPP day-ahead decisions.

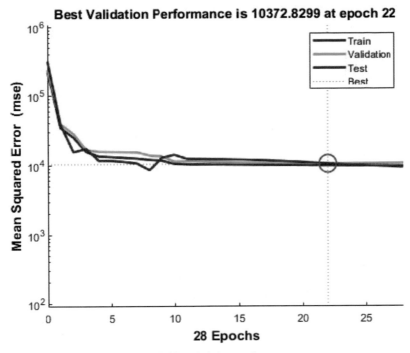

Figure 8.10   Training performance.

8.5 Optimization Block in MHPS

In this block, the GHI on the test day is predicted using trained *model-based* and *model-free* regression models, and then the processing plants' setpoints are optimized to find the optimal setpoints.

8.5.1 Optimization-Assisted ML of MHPS

In this section, the forecasts which are estimated from the *model-based* and *model-free* regression approaches are given as input to the optimization block to estimate the SPP PV production. To solve the optimization problem, a detailed mathematical model of the process plants' DN is required.

In the case study considered, the distribution network of the industry is modeled as a single-feeder radial distribution grid consisting of $N + 1$ buses and lines connecting these buses is modeled by a tree graph. Let $\mathcal{N}_0 := \{0, \ldots, N\}$ denote set of all nodes and $L := \{1, \ldots, N\}$ denote set of distribution lines. The substation node (root node) is indexed as node 0 that connects to the external transmission network. All the non-substation nodes corresponding to $\mathcal{N} := \{1, \ldots, N\}$ represent the non-substation nodes. For each node $n \in \mathcal{N}_0$, let $v_{n,t}$ denote its squared voltage magnitude at time t and $s_{n,t} = p^c_{n,t} + jq^c_{n,t}$ denote the complex power injected to the node n. Furthermore, for each line $n \in L$, let $z_n = r_n + jx_n$ denote its impedance and $S_{n,t} = P_{n,t} + jQ_{n,t}$ denote the complex power flowing from the sending bus π_n. Also let $\mathcal{C}_n$ denote the set of children nodes corresponding to node n.

It should be noted that, even though the considered industry operates with poly-phase lines and balanced loads, the three-phase lines are converted to single-phase using their positive sequence impedance adopting the method in [27].

Modeling of Solar Power Plant:

Consider the SPP plant connected to node n with active power $p^{\text{pv}}_{n,t}$. Let d^{pv}_n, A^{pv}_n, η^{pv}_n, and $G^{\text{pv}}_{n,t}$, respectively, denote the panel derating factor, panel area, panel efficiency, and solar irradiance of the SPP. Due to the intermittency of solar irradiance, it should be noted that the solar irradiance $\widehat{G}^{\text{pv}}_t$ is estimated using *model-based* and *model-free* methods as described in the preceding sections rather than assuming a perfect forecast. Additionally, in this work, it is assumed that the reactive power capability of the SPP is absent, implying that SPP can only support active power. Since the high penetration of $p^{\text{pv}}_{n,t}$ complicates the operation of captive power plant, the excess PV power can be curtailed which is denoted by $\alpha^{\text{pv}}_{n,t}$. The following equations, i.e., Equation (8.7a)–(8.7d), describe the operation of SPP:

$$p^{\text{pv}}_{n,t} = d^{\text{pv}}_n A^{\text{pv}}_n \eta^{\text{pv}}_n \hat{G}^{\text{pv}}_{n,t} \tag{8.7a}$$

$$onp^{\text{pv}}_{n,\min} \leq p^{\text{pv}}_{n,t} \leq onp^{\text{pv}}_{n,\max} \tag{8.7b}$$

$$p^{\text{pv}}_{n,t} = \left(1 - \alpha^{\text{pv}}_{n,t}\right) p^{\text{pv}}_{n,t} \tag{8.7c}$$

$$0 \leq \alpha^{\text{pv}}_{n,t} \leq 0.3 \forall n \in \mathcal{N}. \tag{8.7d}$$

In this work, the cost for operating SPP plant is given by $\Gamma^{\text{PV}}_n = k^{\text{pv}} p^{\text{pv}}_{n,t}$, where $k^{\text{pv}} = \$0.1/\text{kWh}$.

Modeling of Captive Power Plant:

The equations governing the operation of CPP are given as follows:

$$p^{\text{CPP}}_{n,\min} \leq p^{\text{CPP}}_{n,t} \leq p^{\text{CPP}}_{n,\max} \tag{8.8a}$$

$$q_{n,\min}^{\mathrm{CPP}} \leq q_{n,t}^{\mathrm{CPP}} \leq q_{n,\max}^{\mathrm{CPP}} \tag{8.8b}$$

where $p_{n,t}^{\mathrm{CPP}}$ and $q_{n,t}^{\mathrm{CPP}}$, respectively, denote the active and reactive powers of the CPP. The cost for operating CPP is given by $\Gamma_n^{\mathrm{CPP}} = k_n^{\mathrm{CPP}} p_{n,t}^{\mathrm{CPP}}$ with $k_n^{\mathrm{CPP}} = \$0.86/\mathrm{kWh}$. Since the efficiency of the CPP varies with the p_t^{CPP}, this time-varying efficiency is not modeled in this work. However, we gauge the efficiency of the CPP by the optimal p_t^{CPP}, i.e., when the optimal value of the CPP is close to its maximum value, then the CPP operates at the peak efficiency.

Furthermore, let $p_{n,t}^{\mathrm{BL}}, p_{n,t}^{\mathrm{L}}, q_{n,t}^{\mathrm{BL}}$, and $q_{n,t}^{\mathrm{L}}$ denote the active and reactive power baseloads and loads of the processing plant.

The focus then shifts to branch-flow modeling of the processing plant's DN. Because power flow equations are nonlinear, the LinDistFlow approximation of power flow equations [28] are employed to solve the power flow equations provided as follows:

$$P_{n,t} = \sum_{j \in \mathcal{C}_J} P_{j,t} - p_{n,t}^{\mathrm{c}} \forall n \in \mathcal{N}_0 \tag{8.9a}$$

$$Q_{n,t} = \sum_{j \in \mathcal{C}_J} Q_{j,t} - q_{n,t}^{\mathrm{c}} - q_n^{\mathrm{sh}} v_{n,t} \forall n \in \mathcal{N}_0 \tag{8.9b}$$

$$v_{n,t} = v_{\pi_{n,t}} - 2\mathrm{Re}\left[z_n^* S_{n,t}\right] n \in \mathcal{N} \tag{8.9c}$$

where q_n^{sh} is the reactive power injected by the capacitor bank connected to node n. Furthermore, define vectors $p_t = \left[p_{1,t}^{\mathrm{c}}, \ldots, p_{N_{c,t}}^{\mathrm{c}}\right]^{T}$, $q_t = \left[q_{1,t}^{\mathrm{c}}, \ldots, q_{N_{c,t}}^{\mathrm{c}}\right]^{T}$ denote the net active and reactive power consumption. Also, let $v_t = [v_{1,t} \ldots, v_{N_t}]^{T}$ collect the squared voltage magnitude for all nodes $n \in \mathcal{N}$. By adopting [29], the squared voltage magnitude relates to the net active and reactive power injections as given in the following equation:

$$v_t = Rp_{(t)} + Xq_{(t)} + 1_{NT} v_0 \tag{8.9d}$$

where $R = 2F\mathrm{diag}\,(r)\,F^{\tau}$, $X = 2F\mathrm{diag}\,(\mathrm{x})\,F^{\tau}$. Equation (8.9d) relates power injections $\mathbf{p}_t$ and $\mathbf{q}_t$ to the squared voltage magnitude. If the entries of $\mathbf{r}$ and $\mathbf{x}$ are positive, then the matrices $\mathbf{R}$ and $\mathbf{X}$ are symmetric and positive definite. Also, at each time period t, the nodal voltages should satisfy the limits dictated by ANSI C.84.1 given as in the following equation:

$$v_{\min} \leq v_t \leq v_{\max}. \tag{8.9e}$$

Since the CPP plant operates in the grid-connected mode, the bounds on active and reactive powers supplied by the utility grid are given as Equation (8.9f)–(8.9g)

$$p_{min}^{UG} \leq p_t^{UG} \leq p_{max}^{UG} \tag{8.9f}$$

$$q_{min}^{UG} \leq q_t^{UG} \leq q_{max}^{UG}. \tag{8.9g}$$

The cost of the active power imported from the utility grid is given as $\Gamma_t^{UG} = k^{UG}p_t^{UG}$, where $k^{UG} = \$0.11/kWh$.

The optimization problem for process plant can be stated as follows:

$$(P3) \quad \min_{\{p_t, q_t, v_t, \alpha_t\}} \left(\Gamma^{PV} + \Gamma^{CPP} + \Gamma^{UG}\right)$$

$$\text{s.t.} \quad (8.7a) - (8.9g)$$

The problem (P3) is a *linear programming* problem because the objective and constraints are linear and can be easily solved with off-the-shelf linear programming solvers with guaranteed *global optima*. It is worth mentioning that, while the linear approximation of the AC power flow is used to solve the problem (P3), the optimal setpoints are validated using the nonlinear solution ETAP, i.e., when the optimal setpoints are fed into the ETAP solver, the voltage magnitude accuracy of the LinDistFlow and ETAP equations is compared, with the LinDistFlow equations overestimating the actual non-linear voltages. If the voltage magnitudes computed from the ETAP exceed Equation (8.9e), the lower or the upper bound of Equation (8.9e) is tightened further to improve approximation. Algorithm 1, as shown in the box below, gives the steps to implement the proposed.

Algorithm 1: Algorithm to implement the proposed approach
Input: Fetch the load quantities and operational constraints from the processing plant $\{p^L, q^L, p^{BL}, q^{BL}, v_{min}, v_{max}, p_{min}^{UG}, q_{min}^{UG}, p_{max}^{UG}, q_{max}^{UG}, p_{min}^{CPP}, q_{min}^{CPP}, \mathbf{R}, \mathbf{X}\}$
Outputs: $\alpha^{PV}, \mathbf{p}^{UG}, \mathbf{q}^{UG}, \mathbf{p}^{CPP}, \mathbf{q}^{CPP}, \mathbf{p}^{PV}$
Step 1: Train the *model-based* and *model-free* regression models using previous inputs and get optimal β^* and $(\theta^*, \mathbf{b}^*)$
Step 2: Predict the test day GHI using the trained models
Step 3: Perform optimization (P3) $$\min_{\{p_t, q_t, v_t, \alpha_t\}} \left(\Gamma^{PV} + \Gamma^{CPP} + \Gamma^{UG}\right)$$ $$\text{s.t.} \quad (8.7a) - (8.9g)$$
Step 4: Fetch the optimal outputs
Step 5: Validate with ETAP and compare models

Table 8.5   Industry technical parameters.

Parameter	Nominal	Minimum	Maximum
p^{CPP} (MW)	16	16	18
q^{CPP} (MVAR)	12	0	13.5
p^{UG} (MW)	0.5	-5	7
q^{UG} (MVAR)	2	0.3	6
p^{PV} (MW)	-	0	6
p^{L} (MW)	15	0	15
q^{L} (MVAR)	12	0	12
p^{BL} (MW)	10	10	10
q^{BL} (MVAR)	2.25	2.25	2.25
r_1 (Ω)	0.081	-	-
r_2 (Ω)	0.0027	-	-
x_1 (Ω)	0.1165	-	-
x_2 (Ω)	0.0188	-	-
$v_1 \, (\mathrm{pu})^2$	1	-	-
$v_2 \, (\mathrm{pu})^2$	1	$(0.91)^2$	$(1.05)^2$
$v_3 \, (\mathrm{pu})^2$	1	$(0.91)^2$	$(1.05)^2$
α^{PV}	-	0	0.3
q^{sh} (MVAR)	2.3		
$d^{\mathrm{PV}}, A^{\mathrm{PV}}, \eta^{\mathrm{PV}}$	(0.85, 35295, 0.155)		

8.5.2 Experimental Setup

The details of real-time operating conditions with constraints, decision variables, and objective functions of the case study industry to solve problem (P3) are given in Table 8.5. Furthermore, for simulations in this work, a laptop with 11th Gen Intel, core i5-1135G7 at 2.40 GHz, 16.0-GB RAM, and a 64-bit operating system, x64-based processor, is used with MATLAB 2018a. The load profile of the considered industry is constant because the load on the process plant is constant. The load, PV, and net power injection of the industry are shown in Figure 8.11. It can be seen that the net power injections from the utility grid resembles the duck curve [30]; this is due to excessive PV penetration in the afternoon and may result in the overvoltage issues and deteriorate the efficiency of the CPP. The optimization problem (P3) is solved with time granularity of 1 hour.

8.5.3 Validation Block

Using the actual nonlinear power flow solver ETAP, this block validates the optimized setpoints computed in the previous stage. This block aims to examine the effects of optimized setpoints calculated from estimated GHI values,

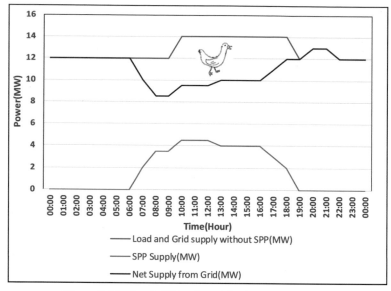

Figure 8.11   Load, PV, and net injections from the UG.

specifically the nodal voltages of the process plant, which are computed using ETAP to see if the voltage magnitudes obey their respective lower and upper bounds. The nodal voltage magnitudes should be well within the range for the process plant to operate satisfactorily; otherwise, technical issues may arise, resulting in abnormal operation. Specifically, the plants' turbo generators and SPP may trip if the voltage falls below the specified limits and results in the complete production loss.

8.5.3.1 Thorough comparisons in voltage-magnitudes for the actual test day for model-based and model-free approaches

Here, the optimization problem is repeated to compute the optimal setpoints for the test day considering the predictions from *model-based* and *model-free* approaches and the corresponding voltage magnitudes are computed from the ETAP.

The optimal inputs computed from the GHI prediction from *model-based* and *model-free* regression approaches, as well as the true field GHI value (benchmark), were used to solve the ETAP power flow. The power flow in the ETAP for the model-based regression approach is depicted in Figure 8.12 for a single snapshot. The voltage magnitude accuracy between the LinDistFlow

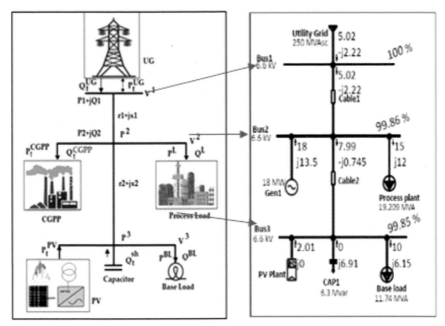

Figure 8.12 SLD of industry (left) and power flow result with ETAP (right).

and the ETAP are below 0.02% pu for all optimizations, i.e., *model-based*, *model-free*, and *true-prediction*.

Next, nodal voltage magnitudes computed from the *model-based, model-free*, and *true-prediction* optimization are depicted in Figure 8.13, from which the following conclusions can be drawn.

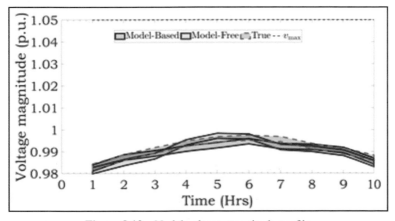

Figure 8.13 Nodal voltage magnitude profile.

Table 8.6 Optimal value comparison.

Sl. no.	$\Gamma^*_{model-based}$	$\Gamma^*_{model-free}$	Γ^*_{true}
1	0.0602	0.0602	0.0601

Key Observations:

The voltage magnitudes of the *model-based* and *model-free* regression approaches are nearly close when compared to the voltage magnitudes of the *true-GHI*, as shown in Figure 8.13. Furthermore, when compared to the *model-based* approach, the *model-free* regression approach performed *slightly worse* in the first 3 hours. However, for the remainder of the time, the *model-free* regression approach closely tracks the *true-prediction*. It is worth noting that both the *model-free* and *model-based* approaches follow their respective true values during the final hours, i.e., hours 8-10. Although, both *model-based* and *model-free* regression models have training R^2 values in the range of 0.76-0.81, the process plants' nodal voltage magnitude is well within the bounds. As a result, it can be concluded that for the specific site conditions, if the process plant loads are fixed, the solar forecasting error can range between $\pm 15\%$. According to the simulations, the majority of the SPP active power generation is exported back to UG, allowing the CPP to operate at full capacity, resulting in improved CPP efficiency. The focus then shifts to the optimal value comparisons.

Optimal Value Comparisons:

In Table 8.6, Γ^* represents the optimal value. According to Table 8.6, even though both *model-based* and *model-free* regression models achieved the same optimal value, this optimal value is less than the *true-predicted* optimal value. However, the difference is negligible. As a result, even though the respective R^2 values of ML models are not close to 1.0, the ML methods can still achieve the lower optimal value. Therefore, the process plants can still be benefited even if the conventional forecasting methods are replaced by the *optimization-assisted* ML methods.

8.6 Conclusion

The ML techniques are used to evaluate the optimal operation of a grid-connected CPP coupled with an SPP. The case study of a well-known industry of a comparable sort is chosen. Forecasting algorithms for solar irradiance are proposed, with promising results. In addition, *optimization-assisted* ML techniques are used to solve the real-time problem. The simulation findings of the proposed approach are satisfactory when validated with ETAP, a standard power system analysis software used in the industry. This work was primarily concerned with the optimal operation of PV integrated co-generation plants with integration concerns. Other integration issues such as reactive power and protective system are not considered and can be addressed in the future work. Future work will also consider load uncertainty and poses the stochastic optimization problem to ensure smooth operation of the process plant. Other pertinent research directions include the use of decentralized control and blockchain technology schemas for peer-to-peer power trading between the industries with SPP and battery energy storage.

References

[1] M. L. Merlin Sajini, S. Suja, and S. Merlin Gilbert, "Impact analysis of time-varying voltage-dependent load models on hybrid DG planning in a radial distribution system using analytical approach," *IET Renewable Power Generation*, vol. 15, pp. 153–172, 2021.

[2] R. H. T. Moger, "Comprehensive review on low voltage ride through capability of wind turbine generators," International Transactions on Electrical Energy Systems, vol. 30, 2020, Art. no. e12524.

[3] D. Chakravorty, J. Meyer, P. Schegner, S. Yanchenko, and M. Schocke, "Impact of modern electronic equipment on the assessment of network harmonic impedance," *IEEE Transactions on Smart Grid*, vol. 8, no. 1, pp. 382–390, Jan. 2017.

[4] B. K. Reddy and A. Kumar Singh, "Reactive power management and protection coordination of distribution network with high solar photo-voltaic penetration," in *Proceedings of 12th International Renewable Engineering Conference (IREC)*, 2021, pp. 1-6.

[5] H. Wang, J. Wang, Z. Piao, X. Meng, C. Sun, G. Yuan, and S. Zhu, "The optimal allocation and operation of an energy storage system with high penetration grid-connected photovoltaic systems," *Sustainability*, vol. 12, no. 15, 2020.

[6] M. Usama, H. Mokhlis, M. Moghavvemi, N. N. Mansor, M. A. Alotaibi, M. A. Azam, and B. A. Akram, "A comprehensive review on protection strategies to mitigate the impact of renewable energy

sources on interconnected distribution networks," *IEEE Access*, vol. 9, pp. 35740–35765, 2021.

[7] G. Shobha and S. Rangaswamy, "Machine learning" in *Handbook of Statistics*, N. G. Venkat and C. R. Rao, Eds. Elsevier, vol. 38, 2018, pp. 197–228, Ch. 8.

[8] A. Mosavi, M. Salimi, S. F. Ardabili, T. Rabczuk, S. Shamshirband, and A. R. Varkonyi-Koczy, "State of the art of machine learning models in energy systems: A systematic review," *Energies*, vol. 12, 2019, Art. no. 1301.

[9] A. Ioannis, R. Valentin, C. Benoit, K. Desen, N. Sonam, K. Aristides, F David, E. G. Sergio Elizondo, and W. Steve, "Artificial intelligence and machine learning approaches to energy demand-side response: A systematic review" *Renewable and Sustainable Energy Reviews*, vol. 130, Sep. 2020.

[10] J. P. Lai, Y. M. Chang, C. H. Chen, and P. F. Pai, "A survey of machine learning models in renewable energy predictions," *Applied Sciences*, vol. 10, no. 17, 2020.

[11] N. Kalfallah, F. Benzergua, I. Cherki, and A. Chaker, "USE of genetic algorithm and particle swarm optimisation methods for the optimal control of the reactive power in Western Algerian power system," *Energy Procedia*, vol. 74, 2015.

[12] H. Khalid and A. Shobole, "Existing developments in adaptive smart grid protection: A review," *Electric Power Systems Research*, vol. 191, 2021.

[13] Y. Xiao, R. Zhao, and W. Wen, "Deep learning for predicting the operation of under-frequency load shedding systems," in *Proceedings of 2019 IEEE Innovative Smart Grid Technologies - Asia (ISGT Asia)*, 2019, pp. 4142-4147.

[14] Z. M. Çınar, A. N. Abdussalam, Q. Zeeshan, O. Korhan, M. Asmael, and B. Safaei, "Machine learning in predictive maintenance towards sustainable smart manufacturing in Industry 4.0," *Sustainability*, vol. 12, no. 19, 2020.

[15] G. Tianhan and L. Wei, "Machine learning toward advanced energy storage devices and systems," *iScience*, vol. 24, no. 1, 2021, Art. no. 101936.

[16] V. T. Nikita, G. K. Victor, N. S. Denis, and V. Z. Alexey, "Machine learning techniques for power system security assessment," *IFAC-Papers Online*, vol. 49, no. 27, pp. 445–450, 2016.

[17] REC-Solar. [Online]. Available: http://www.solardesigntool.com/comp onents/module-panel-solar/REC-Solar/2307/REC255PEBLK/specific ation-data-sheet-html (accessed on 10 April 2020).

[18] IEEE Standard for Interconnection and Interoperability of Distributed Energy Resources with Associated Electric Power Systems Interfaces, *IEEE 1547-2018*.

[19] S. P. Burger, J. D. Jenkins, S. C. Huntington, and I. J. Perez-Arriaga, "Why distributed," *IEEE Power & Energy Magazine*, vol. 17, no. 2, pp. 18–24, Mar./Apr. 2019.

[20] K. Ogimoto and H. Wani, "Making renewables work," *IEEE Power & Energy Magazine*, vol. 18, no. 6, pp. 47–63, Nov./Dec. 2020.

[21] J.-P. Lai, Y.-M. Chang, C.-H. Chen, and P.-F. Pai, "A survey of machine learning models in renewable energy predictions," *Applied Sciences*, vol. 10, no. 17, Aug. 2020, Art. no. 5975.

[22] A. B. Guher, S. Tasdemir, and B. Yaniktep, "Effective estimation of hourly global solar radiation using machine learning algorithms," *International Journal of Photoenergy*, vol. 2020, 2020.

[23] B. Koti Reddy and A. K. Singh "Optimal operation of a photovoltaic integrated captive cogeneration plant with a utility grid using opti-mization and machine learning prediction methods," *Energies*, vol. 14, no. 16, 2021, Art. no. 4935.

[24] B. Facundo, G. Juan, and F. Víctor, "Prediction of wind pressure coeffi-cients on building surfaces using artificial neural networks," *Energy and Buildings*, vol. 158, pp. 1429–1441, Jan. 2018.

[25] A. R. Pazikadin, D. Rifai, K. Ali, M. Z. Malik, A. N. Abdalla, and M. A. Faraj, "Solar irradiance measurement instrumentation and power solar generation forecasting based on artificial neural networks (ANN): A review of five years research trend," *Science of the Total Environment*, vol. 715, 2020, Art. no. 136848.

[26] S. Leholo, P. Owolawi, and K. Akindeji, "Solar energy potential fore-casting and optimization using artificial neural network: South Africa case study," in *Proceedings of 2019 Amity International Conference on Artificial Intelligence (AICAI)*, 2019, pp. 533-536.

[27] W. Kersting, *Distribution System Modeling and Analysis*, 4th ed. 2017. doi: 10.1201/9781420009255

[28] M. E. Baran and F. F. Wu, "Network reconfiguration in distribution systems for loss reduction and load balancing," *IEEE Transactions on Power Delivery*, vol. 4, no. 2, pp. 1401-1407, Apr. 1989.

[29] H. Fontenot, K. S. Ayyagari, B. Dong, N. Gatsis, and A. Taha, "Buildings-to-distribution-network integration for coordinated voltage regulation and building energy management via distributed resource flexibility," *Sustainable Cities and Society*, vol. 69, 2021, Art. no. 102832.

[30] Available: https://www.caiso.com/documents/flexibleresourceshelprene wables_fastfacts.pdf (accessed on 10 July 2021).

9

Establishing a Realistic Shunt Capacitor Bank with a Power System using PSO/ACCS

Ali Mohamed Eltamaly[1,2,3], Osama El Sayed Morsy[4], Amer Nasr Abd Elghaffar[4,5,*], Yehia Sayed Mohamed[5], and Abou-Hashema Ahmed[5]

[1]Electrical Engineering Department, Mansoura University, Egypt
[2]Sustainable Energy Technologies Center, King Saud University, Saudi Arabia
[3]K.A.CARE Energy Research and Innovation Center, Saudi Arabia
[4]Alfanar Engineering Service, Alfanar Company, Saudi Arabia
[5]Electrical Engineering Department. Minia University, Egypt
E-mail: amernasr70@yahoo.com
*Corresponding Author

Abstract

It is important to search for the optimum method to compensate the system reactive power in power systems. The capacitor bank is considered as one of the efficient methods to compensate for the reactive power with a low running cost. Optimal capacitor placement and sizing are key issues to improve the impact of the capacitor bank for enhancing the quality and reliability of the distribution system. Particle swarm optimization (PSO) algorithm has been used in this chapter to select the optimal busbar to add the capacitor and to design the optimal size of this capacitor bank to the distribution network. This chapter discusses the importance of using the shunt capacitor bank for reactive power compensation in terms of improving reliability, loadability, and reduction of power losses. Moreover, this research discusses the ability of automatic capacitor control scheme (ACCS) to add the optimal value online of the shunt capacitors. The proposed method in this chapter has been validated with the IEEE 15-buses power system at 11.0 kV as an example to select the optimal placement and sizing of capacitor banks by using a PSO algorithm automated by the ACCS module for enhancing the quality of the power system.

Keywords: Reactive compensation, capacitor bank, power quality, particle swarm optimization (PSO), automatic capacitor control scheme (ACCS), distribution network.

9.1 Introduction

Power systems around the world are faced with great challenges to keep up with the increasing demand for electric power. Therefore, it is important to ensure the stability, reliability, and efficiency of the power system [1]. The reactive power compensation can improve the power system performance and ensures a high level of stability, reliability, loadability, and efficiency. The main challenge of reactive power compensation is to design the security system with a stable reliability power system [2]. Undervoltage and low/poor power factor in the distribution system will directly affect the system efficiency and increase the power losses. Using the advanced control techniques with the electrical power system can treat the shortage distribution efficiency by compensating the reactive power using shunt capacitor banks, flexible AC transmission system (FACTS) devices with the distribution network [3]. FACTS devices are more accurate and operate fast because it depends on the thyristor for switching, but the FACTS devices are costlier to use with the small-scale distribution network [4]. On the other hand, using the distributed generation (DG) near the load is highly preferred because it can save the generated power from conventional sources and decrease the power losses in long transmission system. But the DG is depending on the renewable energy sources that are directly related to natural environmental causes [5]. Using the capacitor banks parallel with the system under control by the automatic capacitor control scheme (ACCS) techniques can enhance the system voltage by compensating the reactive power. Also, the capacitor bank is very easy to install by using any spare feeder to connect the capacitor bank with the system, which is more economic compared to the active filters or the electric synchronous motors [6]. On further consideration, the capacitor bank installation should be according to the system status and the value of bus reactive power load requirements. Several methods can be used to select

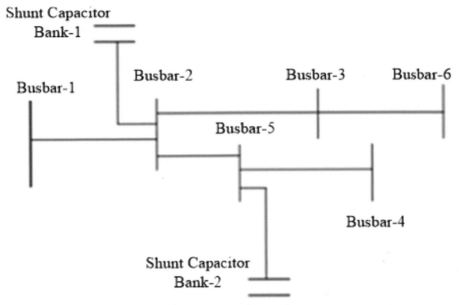

Figure 9.1 SLD shunt capacitor bank for reactive power compensation.

the capacitor allocation, for example, artificial intelligence (AI) [7], genetic algorithm (GA) [8], etc. In this chapter, the particle swarm optimization (PSO) algorithm has been used to optimize the location and size of the capacitor bank in the power system to increase the reliability, stability, and efficiency. PSO algorism has been chosen as an optimization tool because it is very simple, effective, and accurate to catch global optimal solutions [9]. Figure 9.1 shows the single line diagram (SLD) of the shunt capacitor bank for the reactive power compensation. The objective of this chapter can be justified in two main points: the first one is studying the system offline which is simulated by the known stable IEEE-15 bus system with expected load extension which can be solved by the PSO algorithm for a more stable and reliable system, and reduce cost. The results of using the PSO in this chapter can be compared with another research's method to describe the techno-economic asset of using the PSO algorithm. The second contribution in this chapter is the online working of the shunt capacitor bank with the power system, where all researchers study the variant method to find the capacitor bank value and allocation, but it is important to consider the load variant during 24 hours duty time. The load variant is solved in this chapter by using the ACCS new advanced control techniques.

9.2 Problem Statement

Low power factor loads have bad effects on the power quality and the performance of the power systems. Furthermore, the low power factor loads directly increase the power losses, the overheating of feeders/circuits, and minimize the rated transmission power [10]. The capacitor bank draws a leading current and partly the lagging reactive component of the load current which can improve the power factor of the loads [11]. By adding the shunt capacitor bank at the load bus, the power factor will be improved which directly reduces the reactive power consumption cost, increases the rating transmission energy (loadability), and decreases the power losses [12]. Additionally, the power factor correction will improve the overall performance of the power system and will increase the useful life of the loads [13]. The shunt capacitor allocation is considered one of the most important challenges in improving the power system quality. The optimum capacitor allocation will help to improve the voltage profiles and the transmission losses reduction. To determine the best nodes for capacitor placement in the distribution network, it is required to have basic voltage value and power loss reduction data [14]. The application solution methods are depending on the standard capacitor sizes and associated costs. Furthermore, in the objective function determination, the capacitor placement must affect only the flow of reactive power by fixing the real power flow, and the rated voltages are not affected at every point on the feeder; the only balanced three-phase loads are considered. Figure 9.2 shows the variant of electrical loads during 24 hours for three months for Thailand country as an example of the average loads in the world [15]. This example proves that it is not practical to add the permanent fixed capacitor value with the power

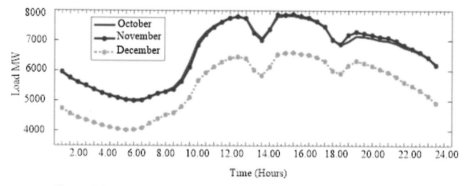

Figure 9.2 Average 24 hours load value during three months of Thailand.

system; so the use of the ACCS is to work with the busbar online to add the required optimal value.

9.2.1 Power Flow Equations

Distribution load flow problems are the main challenges in the power system design. Generally, the ratio R/X_{ratio} in the radial distribution networks is very high; so the power flow problems can be affected by the capacitor placement [16]. Figure 9.3 shows the SLD for the sample distribution network to be used for the power flow calculation. This can be achieved by a set of recursive equations derived from the equivalent current injection at the nth busbar as shown in the following equation [17]:

$$I_n = \left(\frac{S_n}{V_n}\right)^* = \left(\frac{P_n + jQ_n}{V_n}\right)^* \tag{9.1}$$

where I_n is the current at nth busbar, V_n is the voltage at nth busbar, P_n is the active power at nth busbar, and Q_n is the reactive power at nth busbar.

Branch current "B" in the line section between busbars "n" and "$n + 1$" is calculated by the Kirchhoff's current law as

$$B_{n,\,n+1} = I_{n+1} + I_{n+2}.$$

So

$$B_1 = I_2 + I_3 + I_4 + I_5 + I_6, B_2 = I_3 + I_4 + I_5 + I_6$$
$$B_3 = I_4 + I_5 + I_6, B_4 = I_5 + I_6, B_5 = I_6.$$

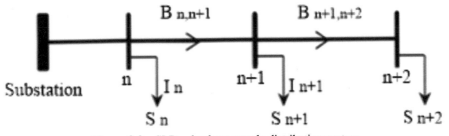

Figure 9.3 SLD n-busbars sample distribution system.

The branch current can be written in generalized matrix form as shown in the following equation:

$$
\begin{bmatrix} B_1 \\ B_2 \\ B_3 \\ B_4 \\ B_5 \end{bmatrix} = \begin{bmatrix} 1 & 1 & 1 & 1 & 1 \\ 0 & 1 & 1 & 1 & 1 \\ 0 & 0 & 1 & 1 & 1 \\ 0 & 0 & 0 & 1 & 1 \\ 0 & 0 & 0 & 0 & 1 \end{bmatrix} * \begin{bmatrix} I_2 \\ I_3 \\ I_4 \\ I_5 \\ I_6 \end{bmatrix}. \tag{9.2}
$$

The matrix can be expressed by considering the bus injection current in the branches as BIBC to be as shown in the following equation:

$$
[B] = [\text{BIBC}] * [I]. \tag{9.3}
$$

The relationship between the branch currents and the busbar voltage can be expressed as shown in the following equation:

$$
V_{n+1} = V_n - B_{n+1}(r_{n,\,n+1} + jx_{n,\,n+1}). \tag{9.4}
$$

So, the power loss in the line section between busbars "n" and "$n + 1$" can be derived as follows [17]:

$$
P_{\text{Loss}}n,\, n+1 = \left(\frac{P_{n,n+1}^2 + Q_{n,n+1}^2}{|V_{n+1}|^2} \right) r_{n,n+1} \tag{9.5}
$$

where $P_{n,\,n+1}$ is the active power between branch "n" and branch "$n + 1$," $Q_{n,\,n+1}$ is the reactive power between branch "n" and branch "$n + 1$," and $r_{n,n+1}$ is the resistance between branch "n" and branch "$n + 1$."

Finally, the total power losses in the system can be defined by the formula shown in the following equation:

$$
P_{T\text{Loss}} = \sum_{n+1}^{nb} P_{\text{Loss}}n,\, n+1. \tag{9.6}
$$

9.2.2 Mathematical Representation

The mathematical equation for minimizing the sum of the power loss and the capacitor costs should be considered the operational and power balance constraints, as shown in the following equation [18]:

$$
\text{Minimize} \sum_{i=1}^{n} K_{ei} T_i P_{Li} + \sum_{j=1}^{K} K_{Cj} C_j \tag{9.7}
$$

where K_{ei} is the constant cost for the energy load level i, T_i is the duration of load level i, P_{Li} is the power loss at ith load level with corresponding time duration T_i, and C_j is the injection kVAR at the jth node.

The minimization of the objective function shown in Equation (9.7) is subjected to power flow balance expressed as shown in the following equation:

$$F_{(X,y)} = 0. \tag{9.8}$$

where X is the unknown power flow variables and y is the known or specified (independent) variables.

Finally, the limits on voltage magnitude are expressed as follows:

$$V^{\min} \leq V_m \leq V^{\max} \tag{9.9}$$

where $V^{\min}$ is the minimum acceptable voltage, V_m is the voltage at mth node, and $V^{\max}$ is the maximum system voltage (1.0 pu practical value).

9.2.3 Sensitivity Calculations

Using the shunt capacitors bank in the correct locations will have different efficiency. The power losses at different bus are subject to many operating factors [19]. The higher sensitive buses are considered the main reference locations to install the capacitor bank [20]. Equation (9.10) shows the variation of the active power losses due to the reactive power compensation at the busbar distribution system

$$\frac{\partial P_L}{\partial Q_i} = 2 \sum_{j=1}^{K} (\alpha_{ij} Q_j + \beta_{ij} P_j) \tag{9.10}$$

where P_L is the total power losses in the system, Q_j is the reactive power at busbar j, and K is the number of locations

$$\alpha_{ij} = \frac{r_{ij} \mathrm{Cos}\,(\theta_i - \theta_j)}{V_i V_j}, \beta_{ij} = \frac{r_{ij} \mathrm{Sin}\,(\theta_i - \theta_j)}{V_i V_j}$$

where θ_i is the voltage angle at node i, θ_j is the voltage angle at node j, and r_{ij} is the real part of impedance between nodes i and j.

The sensitivity value can be calculated by Equation (9.4) at all nodes. So, the higher sensitive location can be used as the coordinate busbar.

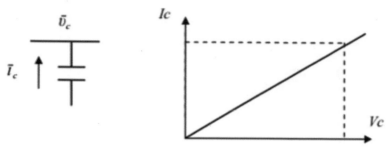

Figure 9.4 Current–voltage characteristic of a capacitor.

9.3 Capacitor Bank Operation Strategies

Using the capacitor bank with the distribution networks can control the reactive power compensation, voltage regulation, power factor correction, and power/energy losses reduction. Planning to select the optimal capacitor allocation and determine the capacitor bank size with the radial distribution system is very important to reach the optimum reactive compensation [21]. The stable voltage system is depending on the optimum selection of the capacitor allocation and size. The main contributions of shunt capacitors with the power system are the flexibility of installation, low cost and operation, a great reduction in power losses, and improvement in voltage levels of weakened busbars. The shunt capacitor bank output characteristic ($I–V$) is a linear rate value between voltage and current as shown in Figure 9.4. The capacitor bank operation with the power system characteristic is defined by Equation (9.11). In the case of transmission systems, shunt capacitors are used to compensate for inductive ($\omega L I^2$) losses and to ensure satisfactory voltage levels during heavy load conditions [22]. Capacitor banks are switched either manually or automatically by voltage relays

$$\overline{V}_C = -jX_C\,(jI_C) = X_C I_C \tag{9.11}$$

where $I_C = \frac{V_C}{X_C} = \omega C V_C$ and $Q_C = \omega C V_C^2$.

Using the shunt capacitor banks with the distribution power system will provide voltage support to a busbar system in case of failure of a tie line or sudden dip in the voltage [23]. Figure 9.5 shows a short transmission system mentioned for the line impedance; the voltage equation in this sample system follows the formulas shown in Equations (9.12)–(9.17):

$$P + jQ = V_r e^{-j\theta}\left[\left(V_s - V_r e^{j\theta}\right)(g_{\mathrm{sr}} + jb_{\mathrm{sr}})\right]$$

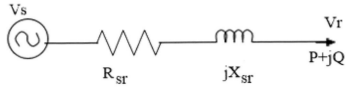

Figure 9.5 A short transmission power system for the power flow indication.

$$= \left[(V_s V_r \cos\theta - V_r^2)\, g_{sr} + V_s V_r b_{sr} \sin\theta\right] +$$
$$+ j\,\left[(V_s V_r \cos\theta - V_r^2)\, b_{sr} - V_s V_r g_{sr} \sin\theta\right] \qquad (9.12)$$

At neglecting the resistance

$$P = V_s V_r b_{sr} \sin\theta \qquad (9.13)$$

$$Q = \left(V_s V_r \cos\theta - V_r^2\right) b_{sr}. \qquad (9.14)$$

Also, at the receiving end, load changes by a factor $\triangle P + \triangle Q$; so

$$\triangle P = (V_s b_{sr} \sin\theta)\,\triangle V + (V_s V_r b_{sr} \sin\theta)\,\triangle\theta \qquad (9.15)$$

and

$$\triangle Q = (V_s \cos\theta - 2V_r)\, b_{sr}\,\triangle V - (V_s V_r b_{sr} \sin\theta)\,\triangle\theta \qquad (9.16)$$

where $\triangle V_r$ is the scalar change in voltage V_r and $\triangle\theta$ is the change in angular displacement. If θ is eliminated from Equation (9.12) and the transmission system resistance is neglected, the dynamic voltage equation of the simulation system is obtained as shown in the following equation:

$$V_r^4 + V_r^2 \left(2QX_{sr} - V_S^2\right) + X_{sr}^2 \left(P^2 + Q^2\right) = 0. \qquad (9.17)$$

As the displacement angle θ is usually very small, from Equation (9.14), we can obtain Equation (9.18) as follows:

$$\frac{\triangle Q}{\triangle V} = \frac{V_s - 2V_r}{X_{sr}}. \qquad (9.18)$$

The receiving end short circuit current can be calculated by the following equation:

$$I_r = \frac{V_s}{X_{sr}}. \qquad (9.19)$$

This assumes that the system resistance is very smaller than the system reactance. So, at no load, $V_s = V_s$; therefore,

$$I_r = -\frac{V_s}{X_{sr}} = -\frac{V_s}{X_{sr}}. \tag{9.20}$$

Thus,

$$\left|\frac{\partial Q}{\partial V}\right| = \text{Short circuit current.} \tag{9.21}$$

Finally, the voltage changes ratio can calculate related to the short circuit level system S_{sc}, as shown in the following equation:

$$\frac{\triangle V_s}{V} \approx \frac{\triangle V_s}{V} \approx \frac{\triangle Q}{S_{sc}}. \tag{9.22}$$

So, Equation (9.21) means the voltage regulation is equal to the ratio of the reactive power change to a short circuit. The obvious results are indicating the receiving end voltage falls with the decrease in system short circuit capacity or increase in system reactance.

9.4 Particle Swarm Optimization

PSO is a metaheuristic evolutionary technique that can be used to reach an effective solution to many engineering problems. The PSO can be extended to absorb the continuous variables in the system [24]. Also, the PSO application can act with the system by considering the evaluation part as a volume-less particle in the dimension search space [25]. PSO method consists of velocity change of each particle at every time step to reach the better individual and global best locations. At every step, the PSO accelerates at random separate term, toward the best global and individual location [26]. Using the PSO algorithm with the population to select the best global (*gbest*) and the best particle [27]. The general engineering optimization problem can be defined as minimizing the $f(x)$.

This general problem can be simulated by Equation (9.22) by considering the particle in d-dimensional space represented is defined by ith

$$x_{i_i} = \{x_{i1},\ x_{i2},\ x_{i3},\ \dots,\ x_{id}\}. \tag{9.23}$$

The best previous position of ith particle is represented as follows:

$$p_{\text{best}_{ii}} = \{p_{\text{best}_{i1}},\ p_{\text{best}_{i2}},\ \dots,\ p_{\text{best}_{id}}\} \tag{9.24}$$

where the change of the position of the particle "*I*" is represented as follows:

$$S_i = \left\{ S_i^1, \; S_i^2, \; S_i^3 \; \ldots, \; S_i^k \right\}. \tag{9.25}$$

The modified velocity and position of each particle can be calculated using the current velocity and the distance from particle best (pbest) to global best (gbest) as follows:

$$S_i^{k+1} = w{*}S_i^k + C_1{*}\mathrm{rand}\,(\,){*}\left(\mathrm{pbest}_i^k - x_i^k\right) + C_2{*}\mathrm{rand}\,(\,){*}\left(\mathrm{gbest}_i^k - x_i^k\right) \tag{9.26}$$

$$S^{\mathrm{min}} \leq S_i \leq S^{\mathrm{max}} \tag{9.27}$$

$$x_i^{k+1} = x_i^k + S_i^{k+1} \tag{9.28}$$

where S_i^k is the velocity of individual constant, pbest is the best value attained by individual i, and gbest is the global best value i at iteration k. C_1, C_2 is the acceleration of the group, and W is the inertia weight factor. rand ( ) is a uniform random number between 0 and 1, and S_{id}^k is the current position of individual i at iteration k.

In the above procedures, the S^{max} parameters can be used to determine the specified resolution and fitness, which searched for areas between the current location and the target location for a very high value of S^{max}, and also the particles will fly past the good solutions [28]. As the S^{max} is a very low value, it may not explore sufficiently beyond local solutions. Considering the S^{max} set from the 10% to 20% range of the dynamic variable, the searching operation becomes stable and reliable.

C_1 and C_2 are the constants of the acceleration terms which pull each particle toward pbest and gbest. Before the particles are tugged back, the result of the low value will allow searching out far of the target zone and the higher results in the movement toward the target regions. So, the settings of the C_1 and C_2 are often 2.0 according to the basic standard [24].

In Equation (9.25), the inertia weight w is the balance between global and local exploration and the exploitation on average results, where w often decreases from 0.9 to 0.4, linearly decreasing as shown in the following equation:

$$w = w_{\mathrm{max}} - \left(\frac{w_{\mathrm{max}} - w_{\mathrm{min}}}{\mathrm{iter}_{\mathrm{max}}} \right) * \mathrm{iter} \tag{9.29}$$

where w_{max} is the maximum value of inertia weight (0.9 in this study), w_{min} is the minimum value of inertia weight (0.4 in this study), $\mathrm{iter}_{\mathrm{max}}$ is the maximum iteration number (generations), and iter is the current iteration number.

9.5 Limitation Treatment

The best particle is found regarding the fitness and converts the inequality constraints to penalty function which added to the best particle. But, in this method, the excellent particle can be rejected regarding the penalty factors. Also, the empirical approach is directly affecting the penalty parameters for solving the problem model. This affection can be avoided by using binary fitness, where one binary fitness is for optimal particle and the other is for the binding constraints.

The optimal particle fitness is the equalized value of Equation (9.25) which indicates the cost of energy losses and the capacitors installed [24–28].

By binding the fitness, it directly extends the violation level, which can be calculated by the following equation:

$$
\text{Binding fitness } (z) = \begin{cases} z_{\min} - z & , & z < z_{\min} \\ z - z_{\max} & , & z > z_{\max} \\ 0 & & \text{else} \end{cases} \tag{9.30}
$$

where z is the inequality value constraint, $z_{\min}$ is the lower limits inequality constraints, and $z_{\max}$ is the upper limits inequality constraints. The binding fitness constraints are considered the first best particle, but if the results are not satisfied, the method will be regenerated. By this method, the particles can be generated, which guarantees to complete the infeasible particles to violate the binding constraints.

9.6 PSO Implementation for Offline Capacitor Study

Figure 9.6 shows the flowchart PSO algorithm for optimal capacitor allocation. The optimum capacitor placement can be defined by the following steps:

1. Defining the candidate allocation:
   - Define the total active and reactive power demand at all busbars in the distribution system.
   - Define the power losses at all branches.
   - Determine the candidate allocation.
   - Find the sensitivity value.
   - Select the higher sensitivity busbars as the optimum candidate allocation.

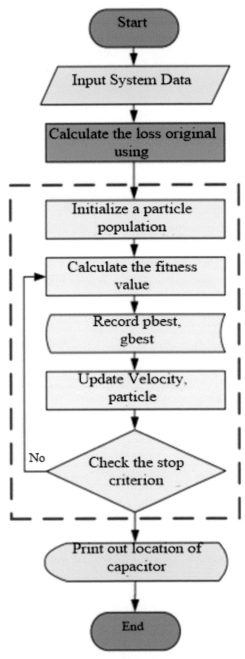

Figure 9.6 Flowchart for optimum capacitor allocation by PSO.

2. Applying data into the PSO algorithm.
   - Input PSO control data.
   - Generate random "n" particle numbers related to every busbar number.
   - Generate the particle velocity as V between the minimum and maximum, where, Vmax = [(maximum capacitor bank rating in MVAR - minimum capacitor bank rating in MVAR)/N number of steps to move the particle from one position to another].
   - Adjust the iter. count = 1.
   - Running the load flows by placing a particle "I" at the candidate busbar for the reactive power compensation.
   - Determine the particle "i" fitness value (base power loss - pl) and comparing the new value with the previous particle best value (pbest).
   - Determine the maximum global among the particle best values.
   - Now, if the new global position is greater than the previous, use the new global position.
   - Using the new pbest and gbest, calculate the velocity Equation (9.25).
   - Set the new velocity value as the prober limit if the new velocity value in limits (-$v$max\, $v$max).
   - Repeat the gbest calculations to find the capacitor values at different load conditions.
   - Calculate the savings obtained with the resultant solution.

9.7 Simulation System for Optimal Capacitor Allocation

The PSO method can use the candidate of the capacitor allocation by determining the required kVAR compensation values of the capacitor bank at varying loads. By assuming "n" is the different load levels and "k" is the number of the points for capacitor allocated, PSO returns "nk" design variables. Meanwhile, the capacitor values cannot find all values in the market. Here, the standard available values in the simulation system, which can be placed in the optimum busbar are 200 and 1200 kVAR. This means, to get 400 kVAR, it is required to duplicate the 200-kVAR value with the optimum busbar system.

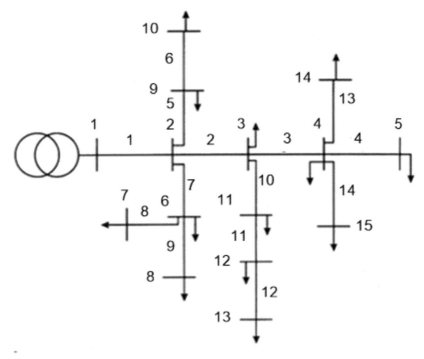

Figure 9.7 Schematic SLD of the IEEE-15-busbar system.

9.7.1 Modified System Data

The benefits of selecting the busbar system to install the shunt capacitor banks are to reduce the power losses and cost. The simulation example used in this part is the IEEE-15 busbar at 11-kV distribution power system. Figure 9.7 shows the SLD for the distribution power system, which presents the busbars data in Table 9.1. The various constants used in the proposed algorithm are Cap_{min} = 200 kVAR, Cap_{max} = 1200 kVAR, K = 0.7259, $C_1 = C_2 =$ 2.05, and w = 1.2. The total loads on the system are (1226.4 + j1251.2) kVA. From the system data as given in Table 9.1, total real losses are 61.79 kW and reactive power losses are 57.30 kVAR. Table 9.2 shows the MW load, MVAR loads, and voltage value for the IEEE-15 busbar simulation system. The simulation IEE-15 busbar loads can be illustrated as shown in Figure 9.8. The increase of the demand power can be reached to 5.02% per year [5], which means the loads in the simulation example will affect the voltage system. Finally, the use of PSO for offline application algorithms can help to decrease the system losses with the optimum capacitor value.

Table 9.1   IEEE-15 busbar simulation system data.

Bus no.	Bus 1	Bus 2	Bus 3	Bus 4	Bus 5	Bus 6	Bus 7	Bus 8	Bus 9	Bus 10	Bus 11	Bus 12	Bus 13	Bus 14	Bus 15
MVA	144.8	76.4	113.7	53.9	112.5	133.7	82.1	109.6	122.2	45.7	138.9	49.5	110.6	66.8	87.5
MVAR	18.9	5.2	21.3	10.8	23.6	14.1	9.3	32.7	20.5	16.1	33.1	7.6	19.5	8.5	13.2
Bus voltage	1.0 009	0.971 283	0.956 669	0.949 952	0.949 918	0.958 232	0.956 008	0.956 954	0.967 971	0.966 897	0.949 952	0.945 829	0.944 517	0.948 608	0.94 844

Table 9.2   General data for IEEE-15 busbar system.

Data	Sending end node	Receiving end node	X (Ohm)	R (Ohm)
Node-1	1	2	1.32349	1.35309
Node-2	2	3	1.14464	1.17024
Node-3	3	4	0.82271	0.841
Node-4	4	5	1.0276	1.52348
Node-5	2	9	1.3579	2.01317
Node-6	9	10	1.1377	1.68671
Node-7	2	6	1.7249	2.55727
Node-8	6	7	0.734	1.0882
Node-9	6	8	0.8441	1.25143
Node-10	3	11	1.2111	1.79553
Node-11	11	12	1.6515	2.44845
Node-12	12	13	1.3579	2.01317
Node-13	4	14	1.5047	2.23081
Node-14	4	15	0.8074	1.19702

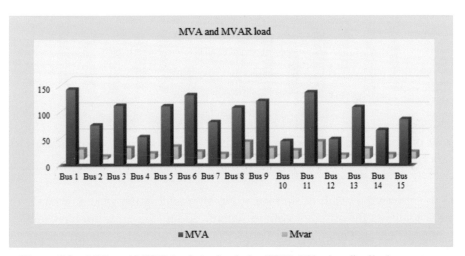

Figure 9.8   MVA and MVAR loads in simulation IEEE-15 busbar distribution system.

9.7.2 Simulation Study

The optimal capacitor bank size at a single location busbar is shown in Figure 9.9 and the corresponding power loss reduction is shown in Figure 9.10, which discusses the highest saving power losses of 23.4 kW.

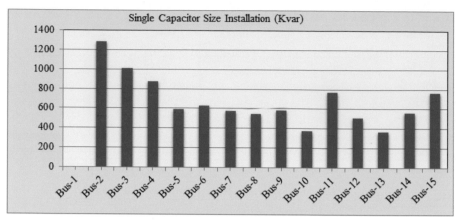

Figure 9.9 Optimal singly capacitor bank value of the 15-busbar system.

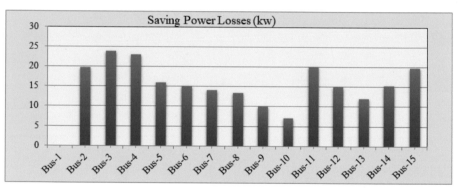

Figure 9.10 Value of saving power losses in the 15-busbar system after using singly capacitor bank.

Table 9.3 shows the result of voltage value after adding the capacitor bank. Figure 9.11 shows the different voltage without adding the capacitor bank and with adding the capacitor bank for IEEE-15 busbars without and with adding the shunt capacitor bank. The power losses can be realized by adding a single capacitor bank in busbar-3 which is 1013 kVAR. Also, at use two busbars for compensation with using busbar-3 compensate system by 1013 kVAR, the second capacitor of 321 kVAR at busbar-6 would provide more saving power losses of 3.7 kW. Also, by adding the capacitor bank in busbar-6, it must remove the capacitor by 142 kVAR from busbar-3 to get a further power losses reduction by 0.5 kW; so busbar-3 was initially over-compensated.

Table 9.3 Voltage value at IEEE-15 busbar simulation system after adding the capacitor bank.

Bus no.	Bus 1	Bus 2	Bus 3	Bus 4	Bus 5	Bus 6	Bus 7	Bus 8	Bus 9	Bus 10	Bus 11	Bus 12	Bus 13	Bus 14	Bus 15
Bus voltage	1.0250	1.0224	1.0069	1.0009	0.99985	1.0086	1.0063	1.007263	1.0188	1.01768	0.9990	0.9955	0.9942	0.9985	0.99835

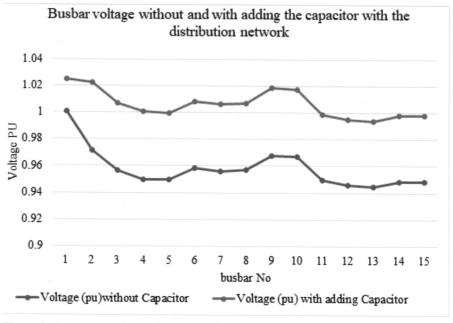

Figure 9.11   Voltage value for IEEE-15 busbars without and with adding the shunt capacitor bank.

But, when both busbar-6 and busbar-3 are compensated, the total power loss reduction will be 32.7 kW by 1192 kVAR of shunt capacitor banks (871 kVAR at busbar-3 and 321 kVAR at busbar-6). To compare these values with the fuzzy-reasoning approach method, which is illustrated in reference [26] and [27], where this method requires adding 1193 kVAR by 805 kVAR at busbar-3 and 388 kVAR at busbar-6. Finally, using the PSO method can save 0.1-kW power losses and it has decreased the capacitor value by 1.0 kVAR.

9.8 Automatic Capacitor Control Scheme

This part aims to discuss the new capacitor control techniques which can contribute to enhancing the power system quality by using the ACCS design. ACCS is the new suggestion system to control the reactive power using the shunt capacitor bank with the distribution network. ACCS design is considered a supervising reactive power technique depending on the intelligent electronic device (IED), which operates to save the stable system continuously by comparing the reference setting values with the actual consumed reactive power value.

9.8.1 ACCS IED Scope

The ACCS/IED modular bay controller unit is considered a programmable logic control (PLC) device, which can be used to control and monitor the switch device. ACCS/IED is designed to operate with the switchgear distribution units equipped with the electrical check-back signaling located in medium-voltage or high-voltage substations [28, 29]. ACCS/IED relay is used to supervise the busbar system in high-voltage substation by the feedback indication to the numerical control relay, which is fed on the transformer or line circuit breaker status. Meanwhile, the ACCS relay contains the integration binary inputs to be connected with the secondary coils to read the voltage and current by the transformer and current transformer, respectively [30]. Also, the ACCS/IED module can absorb more status for measuring power in the power system, by using the measuring transducer which sends the value (MVAR/voltage value) by DC mA DC. ACCS control module is operating manual/automatic, at the releasing interlock for switching the shunt capacitor bank feeder circuit breaker to link with the power system. The ACCS design is considered a monitor continuous technology that improves the action status to improve the power system and the maintenance cost. ACCS/IED can be operated by the remote controller and the system value can be monitored depending on the IEC 61850 communication system with SCADA system [31].

9.8.2 ACCS Operation Logic Steps

Figure 9.12 shows the ACCS operation logic flowchart steps to control the capacitor bank by adding with the system as required and isolate after the system returns back to the normal operation. The main interlock to prevent adding the shunt capacitor bank is the protection healthy condition to avoid the switch on to the fault which can lead to damaging the capacitor bank. Also, the satisfying condition allows adding two capacitor bank units with the system as parallel in one busbar voltage or single unit capacitor, which is used to control the closing time between the two capacitor groups [32, 33]. Using the timer interlock to prevent the close command after satisfying the discharge of the electric charge due to operation. Also, the system reactive power value is continuously compared with the reference set value which initiates to start the timer at the reactive power in the system equal to or greater than the reference value. This timer is used to avoid adding the capacitor bank with the system at the starting high loads, which can lead to overvoltage.

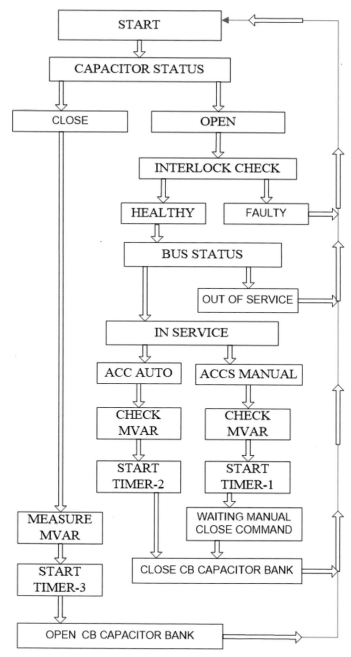

Figure 9.12 Flowchart operation logic steps for ACCS module.

9.8.3 ACCS Operation Sample

Generally, the ACCS/IED is the numerical digital control relay, which is one of the various designers such as SIEMENS, ABB, Schneider, and so on. Using the ACCS/IED to control the reactive power with the power system will able to configure to operate with a multi-stage setting which allows more accurate and flexible operation. Thus, it can control the reactive power by classifying the busbar electric system into two or three parts. Also, during a couple of two parts throw the busbar section to become one busbar system it will digital selecting the busbar system status to select the required optimum configuration. The following example discusses the actual control steps to add the capacitor bank with one busbar, which is classified into two sections with adding 7.1-MVAR capacitor bank with every part. In the sample system, the distribution network is 11-kV voltage, and the selecting busbar system is classified into two sections, which install one capacitor bank unit with every part. The two-section busbar voltage is fed from a separate power transformer 115/11 kV as shown in Figure 9.13, by using the busbar section circuit breaker can couple the two sections to operate

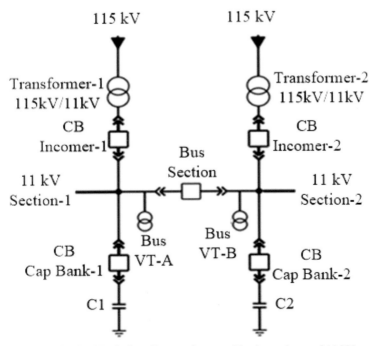

Figure 9.13 Single line diagram for actual busbar voltage of 11 kV.

Table 9.4 Setting values to operate ACCS/IED at different cases.

Quantities	Setting values	Time delay setting	Values
MVAR 1	6 MVAR	T1 – for close C1 and C2 at bus tie CB open	5 minutes
MVAR 2	1 MVAR	T2 – for open C1 and C2 at bus tie CB open	5 minutes
MVAR 3	7 MVAR	T3 – C1 close at bus tie CB close	7 minutes
MVAR 4	2 MVAR	T4 – C1 open at bus tie CB close	7 minutes
MVAR 5	8 MVAR	T5 – C2 close at bus tie CB close	10 minutes
MVAR 6	3 MVAR	T6 – C2 open at bus tie CB close	10 minutes
V1	14.4 kV	T7 open C1 and C2 at OV	5 second

the two power transformers in parallel. Table 9.4 shows the configuration set values to operate the ACCS/IED; the operation logic is considered as the continuous monitor to the actual reactive power in the two sections individually.

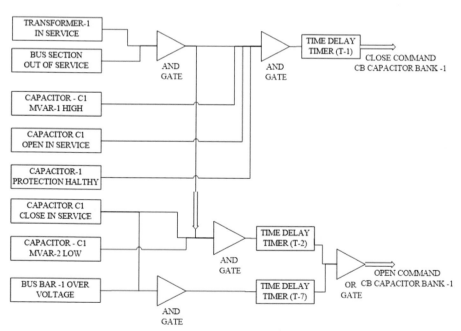

Figure 9.14 Close and open individual logic ACCS/IED module command.

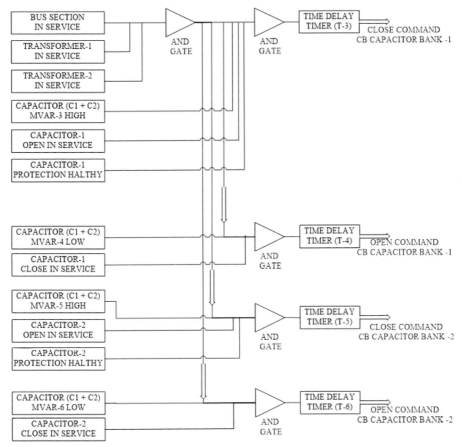

Figure 9.15 ACCS/IED logic for capacitor bank-1 and capacitor bank-2 at the close the bus tie circuit breaker.

For section-1, as the reactive power value reached the setting value (MVAR-2), the ACCS/IED will start to initiate the internal timer-1 to close the capacitor bank-1 (C1) feeder circuit breaker to improve the reactive power with section-1. After that, if the reactive power decreased to the open setting value (MVAR-2), ACCS/IED directly initiates the internal timer-2 to send the open command to the capacitor circuit breaker. At overvoltage in the busbar system which senses through the voltage transformer (Busbar VT-A), ACCS will open the capacitor after timer-7, as shown in Figure 9.14 for the close and open individual logic commands. By the same mirror logic, ACCS/IED can be used to control the adding or isolating of the bus section-2 capacitor bank-2 (C-2), which also detects the undervoltage for section -2 through the voltage transformer (Busbar VT-B). While closing the bus tie circuit breaker, the ACCS/IED module will calculate the reactive power for section-1 and section-2. Figure 9.15 shows the close and open logic commands to the two capacitor bank feeder circuit breakers while closing the bus tie circuit breaker. As the value reaches (MVAR-3), ACCS/IED will start to initiate the internal timer-3; as it reaches (MVAR-4), it will initiate the timer-4 to add the shunt capacitor bank unit-1 with section-1 after timer-3 target. At this time, if the reactive power decreased to be less than the setting (MVAR-4) value, the timer-4 will be released; else it will continue to close the capacitor bank-2 with the system after the timer-4 target. In this case (bus tie circuit breaker closed and capacitor bank-1 and capacitor bank-2 in service), if the reactive power decreases to the open setting value, the ACCS/IED will start to initiate the internal timer-5 and internal timer-6 to open the circuit breaker feeder for the capacitor bank-2 after timer-5 target. After that, if the reactive power is still less than the setting value, it will open the circuit breaker feeder for the capacitor bank-1 after timer-6 target; else it will be released to keep the shunt capacitor bank-1 in service.

9.9 Conclusion

Due to the power system extension and the increase in the reactive loads, the power system has become more sensitive and critical. This extension will influence the distribution network in finding some non-preferred conditions as undervoltage, sag, underfrequency, and so on. In recent years, the reactive power compensation devices have been increased and improved to reach optimum reactive compensation at a low cost. The main reason for using the shunt capacitor bank is present in the power utilities, large industrial to maintain voltage profile at acceptable levels, and to compensate reactive power in distribution and industrial networks, wherever needed and quick amortization of investment are some of the reasons why the using of shunt capacitor banks, which represent an essential reactive power compensation element with the distribution power system. This chapter presented the optimal capacitor bank allocation in the distribution systems by using the PSO method. The approached method in this chapter has been obtained with IEEE-15 busbar voltage as a sample power system, which investigated to provide a total reduction power loss of 32.7 kW by using 1192 kVAR of capacitors by (871 kVAR at busbar-3 and 321 kVAR at busbar-6). Also, this chapter discussed the ACCS to add the shunt capacitor bank with the system to improve voltage profile and reduce the active power loss by adding the accurate required value. ACCS/IED control module simulated and discussed in this chapter for the busbar voltage which classified to two sections, which showed the operation steps at individual sections and at the close the bus tie circuit breaker to be one part. Finally, this chapter aims to recommend the power system designer and the researchers to use the PSO for offline planning and ACCS to add the required capacitor value.

Conflict of Interest

The authors declare that this chapter has no conflict of interest.

Acknowledgement

Dr. Amer Nasr wishes to acknowledge Alfanar Company, especially Mr. Amer Abdullah Alajmi (Vice President, Sales & Marketing, Alfanar Company, Saudi Arabia) and Mr. Osama Morsy (General Manager, Alfanar Engineering Service, Alfanar Company, Saudi Arabia) for their supporting throughout the completion of this research.

References

[1] A. M. Eltamaly and A. N. A. Elghaffar. (November, 2017). Modeling of distance protection logic for out-of-step condition in power system. Electrical Engineering, DOI: 10.1007/s00202-017-0667-3.

[2] A. M. Eltamaly, Y. Sayed, A. H. M. El-Sayed, and A. A. Elghaffar. (2020). Adaptive static synchronous compensation techniques with the transmission system for optimum voltage control. Ain Shams Engineering Journal, DOI: ttps://doi.org/10.1016/j.asej.2019.06.002

[3] A. M. Eltamaly and A. N. A. Elghaffar. (2017). Power flow control for distribution generator in Egypt using facts devices. Acta Bulletin of Engineering Journal, ISSN: 2067–3809.

[4] A. M. Eltamaly and A. N. A. Elghaffar. (August, 2018). Multi-control module static VAR compensation techniques for enhancement of power system quality. Annals of Faculty Engineering Journal. ISSN: 2601–2332.

[5] A. M. Eltamaly and A. N. A. Elghaffar. (December, 2017). Techno-economical study of using nuclear power plants for supporting electrical grid in Arabian Gulf. 2(1), DOI: 10.1007/s40866-017-0031-8.

[6] A. M. Eltamaly, Y. S. Mohamed, A. H. M. El-Sayed, A. N. A. Elghaffar, and A. G. Abo-Khalil. D-STATCOM for distribution network compensation linked with wind generation. In: A. M. Eltamaly, A. Y. Abdelaziz, and A. G. Abo-Khalil (eds). (2021). Control and Operation of Grid-Connected Wind Energy Systems. Green Energy and Technology. Cham: Springer. DOI: https://doi.org/10.1007/978-3-030-64336-2_5.

[7] M. Aman, *et al.* (2014). Optimum shunt capacitor placement in distribution system - A review and comparative study. Journal of Renewable and Sustainable Energy Reviews, 30:429-439, DOI: 10.1016/j.rser.2013.10.002.

[8] M. Mahda, *et al.* (2017). Improve performance in electrical power distribution system by optimal capacitor placement using genetic algorithm. 14th International Conference on Electrical Engineering/Electronics, Computer, Telecommunications and Information Technology, DOI: 10.1109/ECTICon.2017.8096347.

[9] K. Prakash and M. Sydulu. (2007). Particle swarm optimization based capacitor placement on radial distribution systems. Power Engineering Society General Meeting, DOI: 10.1109/PES.2007.386149.

[10] A. Karaarslan, *et al.* (2011). A DSP based power factor correction converter to reduce total harmonic distortion of input current for improvement of power quality. Electrical Engineering Journal, 93(4):247–257, DOI: https://doi.org/10.1007/s00202-011-0215-5.

[11] C. Ersavas, *et al.* (2016). Optimum allocation of FACTS devices under load uncertainty based on penalty functions with genetic algorithm. Electrical Engineering Journal, DOI: 10.1007/s00202-016-0390-5.

[12] A. M. Hemeida, *et al.* (2017). Moth-flame algorithm and loss sensitivity factor for optimal allocation of shunt capacitor banks in radial distribution systems. 19th International Middle East Power Systems Conference (MEPCON), DOI: 10.1109/MEPCON.2017.8301279.

[13] A. M. Eltamaly and A. N. A. Elghaffar. (2018). Enhancement of power system quality using static synchronous compensation (STATCOM). International Journal of Electrical and Computer Engineering, 8(30), EISSN: 2305-0543.

[14] A. A. El-Fergany. (2013). Optimal capacitor allocations using evolutionary algorithms. IET Generation, Transmission & Distribution, DOI: 10.1049/iet-gtd.2012.0661.

[15] P. P. Phyo and C. Jeenanunta. (January-March 2019). Electricity load forecasting using a deep neural network. Engineering and Applied Science Research, 46(1):10-17.

[16] J. H. Teng. (2003). A direct approach for distribution system load flow solutions. IEEE Transactions on Power Delivery, 18(3):882–887, DOI: 10.1109/TPWRD.2003.813818.

[17] N. Gnanase *et al.* (2016). Optimal placement of capacitors in radial distribution system using shark smell optimization algorithm. Ain Shams Engineering Journal, 907-916, DOI: http://dx.doi.org/10.1016/j.asej.20 16.01.006.

[18] S. P. Singh and A. R. Rao. (2012). Optimal allocation of capacitors in distribution systems using particle swarm optimization. Electrical Power and Energy Systems Journal, 43:1267–1275, DOI: http://dx.doi.org/10. 1016/j.ijepes.2012.06.059.

[19] A. M. Eltamaly, M. Y. Sayed, A. H. M. El-Sayed, *et al.* (2020). Power quality and reliability considerations of photovoltaic distributed generation. Technology and Economics of Smart Grids and Sustainable Energy, 5:25, DOI: https://doi.org/10.1007/s40866-020-00096-2.

[20] M. R. Haghifam, *et al.* (2007). Genetic algorithm-based approach for fixed and switchable capacitors placement in distribution systems with uncertainty and time varying loads. IET Generation Transmission, & Distribution, 1(2):244–252.

[21] A. R. Seifi. (2009). A new hybrid optimization method for optimum distribution capacitor planning. Modern Applied Science Journal, 3(4).

[22] S. Corsi, Voltage Control and Protection in Electrical Power Systems, Advances in Industrial Control, Chapter 2 Equipment for Voltage and Reactive Power Control, DOI: 10.1007/978-1-4471-6636-8_2.

[23] J. C. Das. (2015). Power System Harmonics and Passive Filter Designs, Book First Edition, chapter (11), John Wiley & Sons, Inc. IEEE, DOI: 10.1002/9781118887059.ch11.

[24] I. J. Hasan, *et al.*, (2015). Optimal capacitor allocation in distribution system using particle swarm optimization. Applied Mechanics and Materials, 699:770-775, DOI: 10.4028/www.scientific.net/AMM.699.7 70.

[25] M. A. Mohamed, *et al.* (2018). A PSO-based smart grid application for optimum sizing of hybrid renewable energy systems. In Modeling and Simulation of Smart Grid Integrated with Hybrid Renewable Energy Systems, pp. 53–60, DOI: 10.1007/978-3-319-64795-1.

[26] M. H. Haque. (1999). Capacitor placement in radial distribution systems for loss reduction. IEE Proceedings - Generation, Transmission, & Distribution, 146(5), DOI: 10.1049/ip-gtd:19990495.

[27] C.-T. Su and C.-C. Tsai. (1996). A new fuzzy-reasoning approach to optimum capacitor allocation for primary distribution systems. Proceedings of the IEEE International Conference on Industrial Technology.

[28] E. R. Yshi. (1999). Empirical study of particle swarm optimization. Proceedings of the 1999 Congress on Evolutionary Computation. Piscataway, 1945–1950.

[29] M. A. Mohamed, A. M. Eltamaly, and A. I. Alolah. (2017). Swarm intelligence-based optimization of grid-dependent hybrid renewable energy systems. Renewable and Sustainable Energy Reviews, 77: 515–524.

[30] M. Ravindran and V. Kirubakaran. (2012). Electrical energy conservation in automatic power factor correction by embedded system. Energy and Power, 2(4): 51–54, DOI: 10.5923/j.ep.20120204.02.

[31] P. S. R. Murty. (2017). Chapter 17 - Relaying and Protection. Electrical Power System Book. 417–477, DOI: https://doi.org/10.1016/B978-0-0 8-101124-9.00017-6.

[32] Digital Capacitor Bank Control M-6280 Manual. (2008). Digital Capacitor Bank Control for Remote Capacitor Automation, Monitoring and Protection. Beckwith Electric Co. 800-6280-SP-21MC4.

[33] Manual Micom C 264 / C 264C, Bay Computer C264/EN O/C80. Schnider, operation guide.

[34] X.-m. Huang, Y.-j. Zhang, and H.-c. Huang. (2014). Automatic reactive power control in distribution network based on feeder power factor assessment. IEEE PES Asia-Pacific Power and Energy Engineering Conference (APPEEC), ISSN: 2157-4839, DOI: 10.1109/APPEEC.2 014.7066071.

10

Introduction to Blockchain Technologies

Xiaofeng Chen and Xiangjuan Jia

Hangzhou Qulian Technology Co., Ltd. China
E-mail: chenxiaofeng@hyperchain.cn; jiaxiangjuan@hyperchain.cn

Abstract

The energy field is developing rapidly in the direction of a distributed, low-carbon, digital, and intelligent energy Internet. With its characteristics of decentralization, group collaboration, peer-to-peer (P2P) communication, smart contracts, anti-counterfeiting and tamper-proofing, openness, and transparency, blockchain technology fits the development direction of the energy industry to improve quality, efficiency, innovation, and reform. At present, the energy industry of many countries has explored related applications, especially in the business fields of power trading, power metering, microgrid, and renewable energy. As the energy industry pays more attention to energy conservation and energy efficiency, blockchain applications are also attracting attention for optimizing the energy industry.

The International Renewable Energy Agency (IRENA) also lists blockchain technology as one of the key innovations in the field of renewable energy. IRENA pointed out in a report that to accelerate the development of low-cost renewable energy, the world needs at least 30 innovative technology tools and enables them to benefit from the scale of renewable energy. Blockchain technology is one of them.

Blockchain technology is a distributed network data management and data computing technology that uses cryptography technology and distributed consensus protocols to ensure network transmission and access security and realizes multi-party data maintenance, cross-validation, uniformity across the network, and resistance to tampering. As an important evolution of a new generation of information and communication technology, the characteristics of non-tampering, transparent, and traceable data make the blockchain technology an infrastructure to solve the mutual trust of industry chain participants.

Keywords: Blockchain technology, energy, consensus mechanism, smart contract, cryptographic algorithm, core technology, expansion technology, supporting technology.

10.1 Introduction and Classification

The International Renewable Energy Agency (IRENA) said that nearly 200 companies around the world are cooperating with blockchain technology companies[1]. And blockchain-based solutions can support a wider range of energy transactions and grid balancing solutions[2]. Blockchain-based smart contracts can promote the modernization of the power grid and increase the application of renewable energy, especially the intermittent power that is difficult to absorb while reducing costs and speeding up the transaction process[3]. Blockchain technology can realize peer-to-peer (P2P) electricity transactions and can also manage renewable energy and carbon emission reduction certificates and ensure that all transactions cannot be tampered with, making renewable energy transactions and applications more reliable and efficient[4]. So, what is blockchain technology?

Narrowly speaking, blockchain technology is a decentralized shared ledger that combines data blocks in a chain into a specific data structure in chronological order and cryptographically guarantees that it cannot be tampered with or forged. Broadly speaking, blockchain technology is a new decentralized infrastructure and distributed computing paradigm by using encrypted chain block structure to verify and store data, using distributed node consensus algorithm to generate and update data, and using automated script code (smart contract) to program and manipulate data.

There are three types of blockchain technology: public chain, private chain, and consortium chain.

The public chain has the characteristics of open source and anonymity, is free to join and leave, like Bitcoin and Ethereum, and works based on the consensus mechanisms as Proof of Work (PoW), Proof of Stake (PoS), etc. However, anonymous access is currently not suitable for network real-name supervision[5].

A private chain is generally established inside an organization or enterprise, with stronger security and privacy protection capabilities, and can prevent internal and external malicious attacks. However, due to its closed network, it is not suitable for applications that need to be deployed on the Internet and provides services across domains.

Consortium chain is limited to the participation of consortium members. The read and write permissions on the chain and the permission to participate in accounting are set according to the rules of the consortium. The data is limited to the consortium organization and its users to have permission to access or update. Consortium chain has achieved partial decentralization and strong controllability. Practical byzantine fault tolerance (PBFT) and Raft are often used on consortium chains and private chains. These characteristics make the consortium chain have manageable and controllable practical needs and technical support[6].

10.2 Blockchain Technology Characteristics

Blockchain is an innovative application model in the Internet era of computer technologies such as distributed data storage, point-to-point transmission, consensus mechanism, encryption algorithm, etc. Its characteristics are mainly manifested in the following aspects:

10.2.1 Multi-Centralization

From an architectural point of view, the blockchain is based on a P2P network. The damage or loss of any node will not affect the operation of the entire system. The system has excellent robustness. Therefore, the blockchain is a multi-centralized architecture. In terms of storage, the blockchain is a distributed storage technology. Data is distributed and stored in all nodes and a consensus is reached. There is no single center that holds the storage right. Therefore, the blockchain is also a multi-centralized storage[7]. In terms of governance, the blockchain has no centralized organization or institution, and the rights and obligations between any nodes are equal. The blockchain prevents a few people from controlling the entire blockchain system through a consensus mechanism; so the governance of blockchain is decentralized or multi-centralized.

10.2.2 Tamper-Proof, Traceable, and Transparent

Blockchain technology records all transaction behaviors since the creation of the system in the block, which can ensure that the data records on the chain are not susceptible to human tampering, deletion, failure, etc.; so the information exchange activities can be queried and traced. This completely transparent data management system provides reliable tracking shortcuts for operations such as energy transaction processes, logistics tracking, purchase records, use process, and auditing[8].

The blockchain system is open and transparent. In addition to the encryption of the private information of the parties to the transaction, the data is transparent to the entire network nodes[9]. Anyone or any participating node can query the blockchain data records or develop related applications through the public interface. This is the trustworthy foundation of the blockchain system. Blockchain data records and operating rules can be reviewed by the entire network nodes, with high transparency.

10.2.3 High Reliability

From a technical point of view, the essence of blockchain is a distributed database and distributed computing system. Through the form of distributed data storage, each participating node in the blockchain network can obtain a copy of the complete database. Blockchain data is jointly maintained by all nodes, and each node participating in the maintenance can make a copy of a complete record. Therefore, the more the nodes participating in the system and the stronger the computing power, the higher the data security in the system[10].

The database based on blockchain technology is expected to form several global giant databases in the future. All human value exchange activities (including registration, account opening, payment, transaction, clearing, etc.) can be completed in these databases. The business model is highly expandable and inclusive.

10.3 Blockchain Technology Graph

Blockchain is an abbreviation of comprehensive technology. After years of development, it has continuously integrated a variety of today's technologies into one. It is mainly summarized into three categories: core technology, expansion technology, and supporting technology[1]. The core technology refers to the technology that a complete blockchain system must have included, the expansion technology refers to the related technology to further expand the blockchain service capabilities, and the supporting technology refers to the related technology to improve the security of the

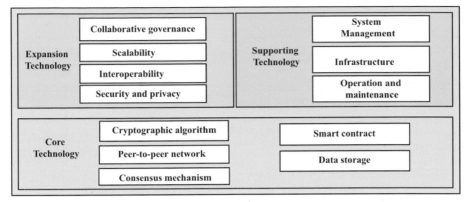

Figure 10.1 Comprehensive graph of blockchain technology.

blockchain system and optimize the user experience. At this stage, blockchain technology consisting of core technology, expansion technology, and supporting technology has gradually formed. In the future, it will continue to innovate and evolve in data circulation, network scale, technical operation and maintenance, and platform security.

Blockchain technology, as shown in Figure 10.1, is a comprehensive technology; its technical composition can be divided into three categories: core technology, expansion technology, and supporting technology according to the degree of importance. Core technology refers to the technology that a complete blockchain system must include, including cryptographic algorithm, P2P network, consensus mechanism, smart contract, and data storage. Expansion technology refers to related technology that further expands blockchain service capabilities, including scalability, interoperability, collaborative governance, security, and privacy. Supporting technology refers to related technology such as improving the security of the blockchain system and optimizing user experience, including system management, infrastructure, operation, and maintenance.

10.3.1 Core Technology Overview

The birth of Ethereum laid the foundation for the five core technologies of the blockchain system, as shown in Figure 10.2. It includes cryptographic algorithm, P2P network, consensus mechanism, smart contract, and data storage.

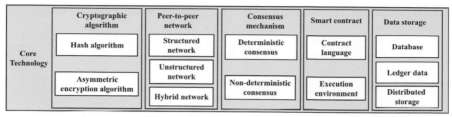

Figure 10.2 Core technology graph of blockchain technology.

(1) Cryptographic Algorithm

Generally speaking, cryptographic algorithms commonly used in blockchain systems include hash algorithms and asymmetric encryption algorithm.

(a) Hash Algorithm

Hash algorithm refers to a method of extracting a fixed-length digital "finger-print" from data of any length. At present, common hash algorithms include MD5 (message digest), SHA-1 (secure hash algorithm), SHA-2, SHA-3, SM3 algorithms, etc.

The hash algorithm has an input and an output. The input is data of any length. Within the algorithm, no matter what form the input data is, it is processed as a simple bit sequence. The output of the hash algorithm is a fixed-length hash value or hash value. Specifically, the hash algorithm scrambles and mixes the data compresses it into a digest, makes the amount of data smaller, and recreates a fingerprint called the hash value. In the blockchain system, the hash algorithm is generally used in the process of transaction verification and block construction, such as building a block. If the data in the block is maliciously added or changed, the resulting hash value will be completely different. In addition, each block header information references the hash value of the previous block, which makes the blocks interlocked and closely connected, thus forming a data chain that is difficult to be tampered with[11].

(b) Asymmetric Encryption Algorithm

An asymmetric encryption algorithm refers to a cryptographic algorithm that uses different ciphertext for encryption and decryption. At present, common asymmetric encryption includes RSA algorithm, Elgamal algorithm, SM2, and elliptic curve digital signature algorithm (ECDSA), of which ECDSA and SM2 are the most used algorithms in the blockchain.

The asymmetric encryption algorithm is composed of a public key and a private key. The public key and the private key are generated by pairing. When the public key is used to encrypt data, only the corresponding private key can be used to decrypt the data; the same applies when the private key encrypts data - only the corresponding public key can be used to decrypt it. Therefore, different keys are used for encryption and decryption, so that its confidentiality will be relatively high, which also solves symmetric encryption. The inconvenience that both parties need to share the key in the algorithm eliminates the need for end-users to exchange keys.

In the blockchain network, each user has a unique pair of public and private keys. The public key is like a bank card number that can be disclosed, and the private key is the non-public part, like the account password corresponding to the bank card. Simply put, in the blockchain, whoever controls the private key has all the data in the account corresponding to the private key[12]. The typical application scenario of asymmetric encryption in the blockchain system is the signature and verification of transactions. The transaction initiator uses the private key to sign the original transaction information and broadcasts the signed transaction and public key. Each node receives after the transaction, the public key can be used to verify whether the transaction is legal. In this process, the transaction initiator does not need to expose its private key, thus achieving the purpose of confidentiality.

(2) Peer-to-Peer Network

The network layer communication protocol of blockchain technology is a P2P network protocol, also known as a P2P network. It is a channel for consensus and information transfer between nodes. It is the network foundation of blockchain technology. It is a non-central server and depends on each node. A network system for exchanging information[13]. Different from a central network service system with a central server, each node in a P2P network can not only act as a requester of network services but also receive requests from other nodes, respond, and provide resources, services, and content. Blockchain technology adopts a P2P network architecture, such as transaction broadcasting, combined with P2P network technology, to decentralize transaction broadcasting to various nodes[14].

Generally speaking, there are three types of P2P networks: structured network, unstructured, and hybrid network.

Structured P2P network node presents a structure of a specific organization, the node supports efficient search of files, and even the searched content is not widely used[15]. In most cases, this is due to the hash function, which facilitates database searches. This is mostly used by consortium chain or private chain which is deployed in private network.

The unstructured P2P network nodes do not have any specific organizational characteristics. Participating nodes randomly communicate with other nodes. Compared with intense liquidity activities (for example, certain nodes frequently join or leave the network), unstructured P2P networks are considered to be robust. This is mostly used by the public chain which is deployed in the Internet.

The hybrid P2P network combines the traditional client-server model with a P2P architecture. For example, it can design a central server that facilitates connections between peers. The interoperability between different blockchains or on-chain and off-chain or blockchains and oracles may meet hybrid networks.

Compared with the other two models, the hybrid model tends to improve the overall performance. The hybrid network model combines the main advantages of the two modes, while simultaneously achieving the two characteristics of high efficiency and distribution.

(3) Consensus Mechanism

The consensus mechanism refers to the process of reaching a unified agreement on the state of the network in a decentralized manner, also known as the consensus algorithm, which is the key to ensuring the consistency of the ledger data of each node on the blockchain platform. As a data structure that stores data in chronological order, blockchain technology can support different consensus algorithms.

During the operation of the blockchain system, the nodes in the distributed network need to use the main chain consensus to reach an agreement on the final blockchain data in the blockchain tree[16]. The main chain consensus refers to the distributed network nodes reaching an agreement on the blockchain data the process of. Regarding whether the blockchain data meets the final consistency, the main chain consensus can be divided into deterministic consensus and non-deterministic consensus[17]. Non-deterministic consensus is also called probabilistic consensus. Deterministic algorithms include Paxos, Raft, and PBFT, and probabilistic consensus algorithms include PoW and departmental PoS[18].

The following is a description of the common consensus algorithms mentioned above.

(a) Probabilistic Consensus

Probabilistic consensus means that the blockchain data reach an agreement with a certain probability, and the probability gradually increases over time. There is no guarantee that blockchain data will not be changed in the future. This has weak consistency. This type is usually applied in the public chain area. This article will introduce the algorithms mentioned above as well as their extended algorithms.

1. PoW:

PoW, the Proof of Work algorithm, which is used in Bitcoin, is an algorithm to determine the contribution of each node according to the rights in the computation process made. The simple understanding is proof to confirm that you have done a certain amount of work. In the blockchain network constructed based on the PoW mechanism, nodes compete for the accounting right by calculating the numerical solution of the random hash, and the ability to obtain the correct numerical solution to generate a block is a specific manifestation of the node's computing power. The PoW mechanism has the advantage of being completely decentralized. In the blockchain with the PoW mechanism as the consensus, nodes can enter and exit freely. However, the computing behavior based on the workload proof mechanism will cause a lot of waste of resources, and the cycle required to reach a consensus is also long[19]. Therefore, PoW is suitable for some resource-based and service-based projects.

2. PoS:

PoS, the Proof of Stake algorithm, is an improved consensus mechanism designed for the shortcomings of the PoW. It is mainly through the method of accounting for rights and interests to solve the problem of inefficiency of the network, waste of resources, and consistency of each node. Different from the PoW, which requires nodes to continuously perform hash calculations to verify the validity of transactions, the PoS adds node weights based on PoW and introduces tokens as the basis for weighting and, based on the proportion and time of each node's weight, reduces the difficulty of PoW in proportion to speed up finding random numbers. The advantage of the PoS is that it solves the problems of waste of resources and low efficiency in PoW, but its decentralization is weaker.

3. DPoS:

DPoS, the deposit-based proof of stake algorithm, is an authorization consensus mechanism similar to the board of directors. This mechanism allows each token holder to vote on the nodes of the entire system, decide which nodes can be trusted and act on their behalf for verification and accounting, and generate a small amount of corresponding rewards.

The decentralization of the blockchain-based on the DPoS mechanism relies on a certain number of representatives, rather than all users. In such a blockchain, all nodes vote to elect a certain number of node representatives, who will act on behalf of all nodes to confirm blocks and maintain the orderly operation of the system. At the same time, all nodes in the blockchain have the power to recall and appoint representatives at any time. If necessary, all nodes can vote to disqualify the current node representatives and re-elect new representatives to achieve real-time democracy[20]. DPoS greatly improves the processing capacity of the blockchain and reduces the maintenance cost of the blockchain, so that the transaction speed is close to that of a centralized settlement system.

(b) Deterministic Consensus

Deterministic consensus refers to the fact that once the blockchain data is agreed upon, it cannot be changed and has strong consistency. This category is usually the traditional distributed consensus algorithm and its improved version.

1. Paxos:

Paxos algorithm is a consensus algorithm based on message passing and is highly fault-tolerant. The problem solved by the Paxos algorithm is how to reach an agreement on a certain value in a distributed system where messages may be delayed, lost, and repeated, to ensure that no matter any of the above exceptions occurs, the consistency of the resolution will not be destroyed.

The Paxos algorithm can be applied to a variety of scenarios. For example, in a distributed database system, if the initial state of each node is the same, and each node performs the same sequence of operations, then they can finally obtain a consistent state. To ensure that each node executes the same sequence of commands, a "consensus algorithm" needs to be executed on each instruction to ensure that the instructions seen by each node are consistent[21].

2. Raft:

Paxos is indeed a very influential consensus algorithm. It can be said that it has laid the foundation for distributed consensus algorithms. However, due to its incomprehension and difficulty in implementation, it is very difficult to implement a complete Paxos algorithm[22]. Therefore, there are many Paxos variants, the most famous of which is the Raft consensus algorithm.

Raft is a distributed consensus algorithm used to manage log replication consistency. Its functions are similar to Paxos, but Raft is easier to understand and implement, and it is easier to apply to actual systems. The difference is that Raft uses stronger assumptions to reduce the state that needs to be considered. It decomposes the consistency problem into three relatively independent sub-problems, namely leader election, log replication, and security. This makes Raft easier to understand and easier to apply to the establishment of actual systems. The Raft algorithm is one of the more consensus algorithms adopted by the alliance chain.

3. PBFT:

PBFT, the practical Byzantine fault tolerance algorithm, is a distributed system consensus algorithm that can tolerate Byzantine errors. The core of PBFT is to form a consensus on the state of the network between normal nodes[23]. It is a consensus mechanism that allows voting and the minority obeys the majority. The algorithm can tolerate Byzantine errors and can allow the participation of strong supervision nodes. The algorithm has high performance and is suitable for the development of enterprise-level platforms[24]. At present, the mainstream enterprise-level blockchain solution Hyperledger Fabric and Hyperchain, which were developed by Hangzhou Qulian Technology, both provide PBFT implementation solutions. However, the original PBFT algorithm is not perfect in terms of reliability and flexibility. Hyperchain has enhanced reliability and flexibility and designed and implemented an improved algorithm of PBFT, namely robust Byzantine fault tolerant (RBFT).

RBFT optimizes the execution process of the native PBFT algorithm, adds dynamic data failure recovery and dynamic node addition and deletion mechanisms, which greatly improves the stability, flexibility, and availability of the system, and better meets the production environment requirements of enterprise-level consortium chain. Realize the single point automatic recovery, automatic recovery strategy of dynamic data of the ledger[24]. This strategy can ensure that the cluster dynamically adds and deletes nodes and dynamic data failure recovery under non-stop conditions greatly enhances the availability of the consensus module and improves the overall transaction throughput and system stability of the system while ensuring the strong consistency of the BFT system performance, reaching ten thousand TPS and millimeter-level delay[25]. RBFT's automatic recovery mechanism makes the storage of its node consistent with the latest storage state in the system as soon as possible by actively requesting the block and the block information that is being consensus. It solves the data recovery problem caused by network jitter, sudden power failure, disk failure, etc., during the operation of the blockchain network, which may cause the execution speed of some nodes to lag behind most nodes or directly go down.

As shown in Table 10.1, the characteristics of common consensus mechanisms are listed.

Table 10.1 Characteristics of common consensus mechanisms.

Algorithm	Minimum node	Fault tolerance	Fraud prevention	Advantage	Disadvantage
PoW	1	The difficulty of algorithm, computing power	<50%	Completely decentralized, Free to enter and exit	Violence causes a lot of waste of resources, and the cycle for reaching consensus is long.
PoS	3	The difficulty of algorithm, computing power, weight	<50%	Reducing the number of nodes participating in verification and accounting can speed up the consensus cycle	Waste of computing resources
DPoS	3	The difficulty of algorithm, computing power, weight	<50%	The number of verification and accounting nodes is further reduced, and consensus verification in seconds can be achieved	Waste of computing resources

Table 10.1 Continued

Algorithm	Minimum node	Fault tolerance	Fraud prevention	Advantage	Disadvantage
Paxos	3	2F+1	F	Efficient, strict mathematical proofs	Hard to understand and implement, crash fault-tolerant
Raft	3	2F+1	F	Easier to understand and implement	Crash fault-tolerant
PBFT	4	3F+1	F	Consensus is confirmation, transaction confirmation time is short, and efficiency is high	The degree of decentralization is not as good as PoW, and the reliability and flexibility are not perfect
RBFT	4	3F+1	F	Supports dynamic node management and failure recovery mechanisms	The degree of decentralization is not as good as PoW

Note: F is the number of crash fault nodes or Byzantine fault nodes.

(4) Smart Contract

The smart contract refers to the custom program logic embedded in the blockchain. It is responsible for implementing, compiling, and deploying the business logic of the blockchain in the form of code, completing the transaction trigger execution of the established rules, and minimizing manual intervention.

According to Turing completeness, smart contracts are divided into two types: Turing complete and non-Turing complete[26]. Turing's complete smart contract has strong adaptability and is suitable for businesses with more complex logic, but there is a risk of falling into an endless loop. Non-Turing complete smart contracts are simpler, more efficient, and safer, but they cannot implement arrays or more complex data structures.

The development of blockchain technology has provided a good operating foundation for smart contracts, and smart contracts have played an important role in the blockchain. The smart contract of the blockchain was first established in Ethereum[27]. Ethereum is an open-source public blockchain platform with smart contract functions. The Ethereum project draws on the technology of the Bitcoin blockchain and expands its application range.

(5) Data Storage

Block is an important data structure in the blockchain ledger, which stores core transaction information. The blocks of the blockchain are linked in a certain order. This is a logical sequence, but it is not necessary to follow this logical sequence when storing but to determine each other through a hash pointer. The logical relationship between the hash pointers is a data structure, to be precise, a pointer to the storage location of the data, and also the hash value of the location data[28]. Compared with ordinary pointers, hash pointers can not only tell you the storage location but also verify that the data has not been tampered with. It is generally stored in the form of the hash value of transaction information. In addition, some important additional information, such as the generation time of the block, the hash value of this block, and the sub-block, is used when nodes compete for the right to keep accounts. Random numbers, etc., these blocks are arranged according to the mutual relationship determined by the hash pointer to form a blockchain data structure, which is the storage in the blockchain.

10.3.2 Expansion Technology Overview

With the continuous expansion of the breadth and depth of blockchain applications, the industry's requirements for blockchain are gradually rising. Changes in industry needs have spawned a series of expansion technologies to optimize the blockchain[29]; as shown in Figure 10.3, it includes four aspects: scalability, interoperability, collaborative governance, security, and privacy.

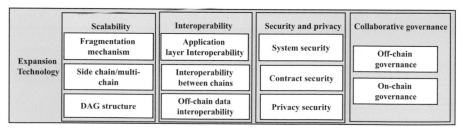

Figure 10.3 Expansion technology graph of blockchain technology.

(1) Scalability

Scalability has become a key technology to improve system performance. Broader scalability includes two aspects: performance scalability and function scalability. Performance scalability focuses on increasing transaction throughput through vertical expansion, and functional scalability focuses on enhancing blockchain service capabilities through horizontal expansion[30].

In Bitcoin, the efficiency of the consensus mechanism is the bottleneck of performance scalability[31]. The overall performance of the whole blockchain system is also impacted by the bandwidth and speed of the P2P network, the performance of sign, verification, a hash of cryptographic algorithms, the single node performance, the IO rate of storage, the execute rate of virtual machine, etc. Therefore, scalability has become a key technology to further enhance the processing capabilities of the blockchain.

At present, common scalability technologies include sharding mechanism, namespace mechanism, lightning network, state channel, directed acyclic graph (DAG) consensus, on-chain, off-chain, side-chain, child-chain, etc.

(2) Interoperability

Interoperability refers to the ability to exchange information between a blockchain system instance and other system instances or components and to use the exchanged information. Among them, other system instances refer to all external system instances except the blockchain system instance. It includes three aspects: application layer interoperability, inter-chain interoperability, off-chain data interoperability.

It refers to the ability to exchange information between the upper application system instance and the underlying blockchain system instance and to use the exchanged information. Specifically, it contains two meanings: (1) data circulation and value sharing between different applications through the underlying chain; (2) docking interaction between the upper-level application system instance and the underlying blockchain system instance.

It refers to the ability to exchange information between different blockchain system instances and use the exchanged information. It can also be called cross-chain. It is mainly manifested in the process of information interaction between different blockchain system instances, including homogeneous chain interoperability and heterogeneous chain interoperability.

It refers to the ability to exchange information between the blockchain system instance and the off-chain data system and to use the exchanged information. It is mainly manifested in the process of secure interaction between the blockchain system and the external data system.

(3) Security and Privacy

Blockchain is a decentralized ledger, and the security protection capabilities of different nodes are uneven, leading to the risk of the system being attacked. Smart contract developers have uneven capabilities, coupled with the lack of convenient and effective smart contract automatic auditing schemes, resulting in smart contract security accidents occurring frequently, and it has become the hardest hit area for blockchain security. The blockchain is decentralized. The characteristics of quasi-anonymity, coupled with the lack of effective supervision methods, lead to the risk of abuse of the chain system. With the continuous enrichment of data on the chain and the continuous expansion of application scenarios, privacy issues in the data circulation process have become increasingly prominent. For example, the smoothness of user identity information, asset information, transaction flow, and other information requires certain technical means to improve the data circulation process and privacy protection capabilities[32]. In addition to traditional data encryption, access control, and data processing, common methods also introduce cryptographic algorithms such as zero-knowledge proofs and homomorphic encryption, as well as TEE, MPC, and federated learning to enhance privacy protection capabilities in multi-party collaboration scenarios. Realize the "available but not visible" data.

(4) Collaborative Governance

Blockchain technology, as a decentralized ledger, emphasizes the model of equal cooperation between all participants. Compared with traditional centralized services, the cooperative model of equal status among multiple parties increases the difficulty of collaborative governance. Blockchain governance refers to the decision-making process of creating, modifying, and updating system rules. It can be divided into two types: off-chain governance and on-chain governance. The common form of off-chain governance is a governance committee formed by core participants, and governance rules are usually voted by the governance committee. On-chain governance means that governance rules are usually encoded in a governance agreement, and each participant makes online voting decisions[32].

10.3.3 Supporting Technology Overview

As a kind of software system, the actual application process of blockchain requires supporting technology to improve system security, optimize user experience, and accelerate the development of blockchain, including system management, infrastructure, operation and maintenance, as shown in Figure 10.4.

(1) System Management

The system management layer is responsible for the management of other parts of the blockchain architecture, which mainly includes two types of functions: authority management and node management. For permission chains, authority management is a key part of blockchain technology, especially for data access. Node management is based on chain-level roles and chain-level configuration, and chain-level administrators conduct node management through proposal voting. Node management includes adding nodes and deleting nodes. As a complex decentralized system that integrates multiple technologies, blockchain is faced with system security and compliance security issues during actual use. Like traditional centralized services, chain system security currently mainly involves network attacks such as DDoS attacks, Sybil attacks, and eclipse attacks. Common defense methods include setting up proxy nodes, strengthening network identity authentication systems, and network current limiting. To promote the stable and compliant development of the blockchain system, it is necessary to strengthen the implementation of the regulatory mechanism in terms of policies, regulations, and technical tools.

(2) Infrastructure

The infrastructure of blockchain technology is divided into two categories: general infrastructure and dedicated infrastructure. General infrastructure refers to the software and hardware resources required during the use of

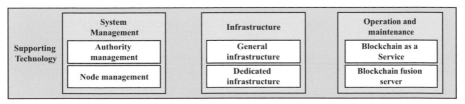

Figure 10.4 Supporting technology graph of blockchain technology.

the chain system and traditional Internet services, and it is versatile, such as communication networks, cloud platforms, etc. Dedicated infrastructure refers to the software and hardware resources specifically required during the use of the chain system, such as unified chain resource management system, and digital identity management system. The general infrastructure has been relatively mature after years of development. Dedicated infrastructure involves the development of blockchain governance and related standards and specifications. The business is in the early stage of the industry. With the eruption of the interconnection needs of different chains, it will promote the construction of blockchain dedicated infrastructure.

(3) Operation and Maintenance

As a complex system that integrates cryptographic algorithms, P2P networks, consensus mechanisms, smart contracts, and other technologies, blockchain is difficult to deploy, operate, and maintain. To lower the barriers to the use of blockchain technology, some blockchain platform applications have emerged, such as blockchain as a service (BaaS) and blockchain fusion server (BFS),and they greatly reduce the difficulty and workload of blockchain system operation and maintenance.

BaaS supports one-stop research and development (R&D) services such as the one-click deployment of alliance chains, visual monitoring operation and maintenance, and smart contract R&D, helping users to focus more on the R&D and innovation of core business and achieve rapid business on-chain.

BFS provides blockchain one-click deployment networking services, presenting the original cumbersome and time-consuming manual deployment process of blockchain networks in an automated and visual way. At the same time, BFS, which is developed by Hangzhou Qulian Technology, provides a full range of operation and maintenance services, through node monitoring visualization, to meet the needs of monitoring data visualization in various scenarios.

For some common operations in operation and maintenance operations, it is recommended that developers formulate a unified operation specification according to their environment, which helps to minimize operation errors and reduce operational risks. Before deployment, a reasonable resource assessment can reduce the frequency of later expansion and enable the system to effectively carry the existing business volume while coping with the subsequent business growth for a certain period. In actual deployment, according to the security of the organization's private key, it can be divided into two scenarios: single-institution deployment and multi-institution P2P deployment, and the deployment can be selected according to the actual scenario. A complete monitoring system can provide early warning in time before the event or provide detailed data after the event, which can be used to track down the problem and effectively shorten the average repair time of anomalies. The monitored indicators mainly include host status, node process, consensus status, error log, and alarm information. Once the deployment is completed and the operation is successful, it will enter the later stage of operation and maintenance. The following operations must be performed in strict accordance with the operation and maintenance specifications, including stress testing, security control, organization management, group management, node management, node upgrade, node migration, certificate management, disk expansion, etc.

10.4 Conclusion

Blockchain technology is born from Bitcoin, which is perfectly integrated with PoW, P2P network, and cryptographic algorithms, and is developed and grown from Ethereum which is a nicely used smart contract. Now that it is stepping into stage 3, where is its development direction?

With the advantages of decentralization or multi-centralization, being tamper-proof, traceability, transparency, and high reliability, blockchain technology will be evolving rapidly from three aspects: core technology, expansion technology, and supporting technology. The innovation of core technologies which include cryptographic algorithms, P2P network consensus mechanism, smart contract, and data storage will improve the performance of blockchain systems. The innovation of expansion technologies which include scalability, interoperability, collaborative governance, security, and privacy will enhance the service capabilities and application breadth of the blockchain system. The innovation of supporting technologies which include system management, infrastructure, operation, and maintenance will improve usability and ease of use of blockchain systems.

As a technology to reconstruct production relations and information infrastructure, fusion with Internet of Things, artificial intelligence, Big data, cloud, and 5G, blockchain technology will be strong enough to face the challenges and risks on compliance, large-scale applications, the real economy, and digital economy.

References

[1] IRENA, *Innovation Landscape for a Renewable-Powered Future: Solutions to Integrate Variable Renewables. Preview for Policy Makers.* Abu Dhabi: International Renewable Energy Agency, 2019.

[2] Hellani, H., Sliman, L., Samhat, A. E., and Exposito, E., On Blockchain Integration with Supply Chain: Overview on Data Transparency. *Logistics*, 2021, *5*(3): 46

[3] IRENA, *Innovation Landscape Brief: Blockchain.* Abu Dhabi: International Renewable Energy Agency, 2019.

[4] Livingston, D., Sivaram, V., Freeman, M., and Fiege, M., Applying Blockchain Technology to Electric Power Systems, 2018.

[5] Idrees, S. M., Nowostawski, M., Jameel, R., and Mourya, A. K. Security Aspects of Blockchain Technology Intended for Industrial Applications. *Electronics*, 2021, *10*(8): 951.

[6] China Academy of Information and Communications Technology (CAICT), Blockchain White Paper, 2020. Available: http://www.trustedblockchain.cn/#/result/result/resultDetail/3b1928a5c5404eaf80958db5f39bef21/0

[7] China Academy of Information and Communications Technology (CAICT), Blockchain Interoperability White Paper, 2020. Available: http://www.trustedblockchain.cn/-/result/result/resultDetail/c90ee957aaf74f20960813999b7ea6bd/0

[8] Musleh, A. S, Yao, G. and Muyeen, S. M., Blockchain Application in Smart Grid-Review and Frameworks. *IEEE Access*, 2019, *7*: 86746-86757.

[9] Blockchain Technology [EB/OL]. [Accessed: 2020-04-03]. Available: https://www.tenet.eu/our-key-tasks/innovations/blockchain-technology.

[10] Yli-Huumo, J., Ko, D., Choi, S., *et al.*, Where is Current Research on Blockchain Technology-A Systematic Review. *PLoS One*, 2016, *11*(10): e0163477.

[11] Castro, M. and Liskov, B. Practical Byzantine Fault Tolerance and Proactive Recovery. *ACM Transactions on Computer Systems (TOCS)*, 2002, *20*(4): 398-461.

[12] Castro, M. and Liskov, B., Practical Byzantine Fault Tolerance. In Proceedings of the Third Symposium on Operating Systems Design and Implementation, 1999, 173-186.

[13] Available: https://en.wikipedia.org/wiki/Quorum _ (distributed - computing)

[14] Schneider, F. B., Implementing Fault-Tolerant Services Using the State Machine Approach: A Tutorial. *ACM Computing Surveys*, 1990, *22*(4): 299-319.

[15] Musleh, A. S., Yao, G., and Muyeen, S. M., Blockchain Application in Smart Grid-Review and Frameworks. *IEEE Access*, 2019, *7*: 86746-86757.

[16] Swan, M., *Blockchain: Blueprint for a New Economy*. USA: O'Reilly, 2015.

[17] Kushch, S. and Castrillo, F. P., A Review of the Applications of the Blockchain Technology in Smart Devices and Distributed Renewable Energy Grids. *ADCAIJ: Advances in Distributed Computing and Artificial Intelligence Journal*, *6*(3): 75-84.

[18] Eranga, B., *et al.*, A Blockchain Empowered and Privacy Preserving Digital Contact Tracing Platform. *Information Processing & Management*, 2021, *58*(4): 102572-102572.

[19] A Brief Introduction to Several Consensus Mechanisms of Blockchain - Katastros.

[20] ISO/TC 307 Blockchain and Distributed Ledger Technologies. [EB/OL]. Available: https://www.iso.org/committee/6266604.html.

[21] Available: http://www.jos.org.cn/html/2021/2/6150.htm#outline_anchor_24.

[22] IEEE Xplore. [EB/OL]. Available: https://sagroups.ieee.org/cts-bsc/.

[23] IEEE P2418.5 - Standard for Blockchain in Energy. [EB/OL]. Available: https://standards.ieee.org/project/2418_5.html.

[24] Available: https://www.hyperchain.cn/products/hyperchain.

[25] Available: http://kns.cnki.net/kcms/detail/32.1117.TV.20210412.1643.006.html.

[26] Cheng, C.-y., *et al.*, Vested Interests: Examining the Political Obstacles to Power Sector Reform in Twenty Indian States. *Energy Research & Social Science*, 2020, *70*: 101766.

[27] Elliptic, The Global Standard for Blockchain Intelligence. [EB/OL]. [Accessed: 2017-06-10]. Available: http://www.elliptic.co/

[28] Blockchain Group, Blockchain Intelligence Group. [EB/OL]. [Accessed: 2017-06-10]. Available: http://Blockchaingroup.io/.

[29] Diestelmeier, L., Regulating for Blockchain Technology in the Electricity Sector: Sharing Electricity – and Opening Pandora's Box? [EB/OL]. [Accessed: 2017-05-09].

[30] Dorri, A., Kanhere, S. S., and Jurdak, R., Blockchain in Internet of Things: Challenges and Solutions. *arXiv preprint arXiv:1608.05187*, 2016.

[31] Cachin, C., Architecture of the Hyperledger Blockchain Fabric. In *Proceedings of the Workshop on Distributed Cryptocurrencies and Consensus Ledger*, 2016.

[32] Available: https://mp.weixin.qq.com/s/fjscjSNZf-kkqX97TBrNvQ.

11

Blockchain Technologies for Renewable Energy Resources with Case Study: SHA–256, 384, and 512

Kaung Si Thu*, Shubham Tiwari, and Weerakorn Ongsakul

Department of Energy, Environment, and Climate, School of Environment, Resources and Development, Asian Institute of Technology, Thailand
E-mail: a.kaungsithu@outlook.com
*Corresponding Author

Abstract

Blockchain is a new technology that has captured the attention of energy companies, supply companies, entrepreneurs, software companies, investment firms, government bodies, and academics. This chapter analyzes the performance of the hardware used to simulate the peer-to-peer energy trading between distributed generations with proof-of-work (PoW) consensus mechanism especially in the case of SHA-256, 384, and 512. The PoW mechanism creates a blockchain environment to trade between parties without authorization, and it establishes a systematic methodology for the trading of energy with an advanced contract system. Performance is measured by the CPU and GPU of the device during the complete period of simulation. Linear trendline and percentage evaluation demonstrate different scenarios of three secure hash algorithms (SHAs). The result shows that SHA-512 has the highest performance efficiency in terms of hardware usage and transaction duration in peer-to-peer energy trading systems. It means that SHA-512 takes more memory size in creating a block than other SHAs; however, it has sustainable development in the processing of the blockchain demonstration.

Keywords: Renewable energy, blockchain, secure hash algorithms (SHAs), peer-to-peer, energy trading.

11.1 Introduction

Energy is an absolute necessity for developing a sustainable society. The trading of energy is a vertical approach from generation to end-user; however, technology advancements and market behavior have led to the inclusion of customers into the electricity trading market, where prosumers can trade their excessive amount of energy [14, 15]. This type of market can reduce carbon emissions, generate profit for both prosumers and consumers, and benefit the grid system without absolute centralized authorization. The decentralized energy system is used to mitigate the information, communication, and technology in every aspect of the energy market to solve sustainable development goals, energy efficiency, security, and optimization. Since diverse energy sources exist, the market might be turbulent at any time [11–13].

The integration of blockchain is widely acknowledged as a paradigm shift to a decentralized energy market with bi-directional power flow [16]. When merged with smart contracts, it offers accessible, tamper-proof, and reliable platforms that can allow technology solutions. This research allows us to demonstrate the hardware performance in different usage of secure hash algorithms (SHAs) at peer-to-peer energy trading between distributed generations with a consensus mechanism. The proof-of-work (PoW) mechanism creates a blockchain environment to trade between parties without authorization, and it will establish a systematic methodology for the trading of energy with an advanced contract system. The hypothesis is that transaction time in the PoW mechanism with less computation power depends on the different use of SHAs in the trading system. Additionally, the performance comparison is evaluated on distinct SHAs used in the consensus mechanism-based trading [17, 20–22].

11.2 Local Energy Trading and Consensus Algorithms

In the event of local energy trading, the community organizations will profit as well as the utility grid infrastructure will be less stressed [7]. When it comes to the energy market, the local grid can function as an ideal client by supplying and selling electricity between grids [8]. While the local grid systems have been generally approved after the feasibility pilot project, the cost of installing renewable energy sources and storage technologies is still a barrier that must be solved.

This research used the solar generation and consumption data from an institute at Thailand – Asian Institute of Technology. Installed with solar panels at Library and Energy department buildings, they produce electricity during the daytime and consume, which acts as prosumer models. This study compares the performance of blockchain simulation under various SHAs (SHA-256, SHA-384, and SHA-512) based on the author's previous research [10]. Moreover, the research [9] demonstrated to promote the potential local energy trading within the institute if more participation is involved, and [10] proved the laboratory simulation of blockchain-based energy trading in prosumer bodies.

Demand-side integration is a critical component of dispersed generating services, and it may be classified based on how demand fluctuations are generated by different factors. However, decentralized consortium blockchains are spread across a variety of hardware owned by different entities [23–25]. A ledger system or blockchain database, a chain of blocks network that reflects transactions. Using a blockchain, people can send information directly to one another, creating a link between them and respective data. Each client joins the network using their blockchain application node. The identification of a client is determined by a cryptographic key pair. A smart contract is a protocol, necessarily for blockchain-based energy trading that optimizes the execution of the corporate activity, and the right procedure is done by

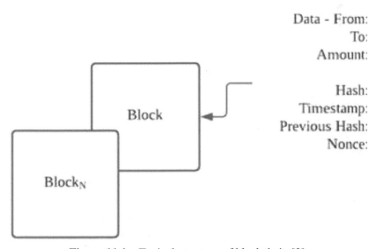

Figure 11.1 Typical structure of blockchain [3].

the consensus mechanism. A cryptographic key is extracted, allowing for accessing the user identification, which is an address that is unique to each user's server. A block typically has information of index, timestamp, nonce, transaction data, a hash function of the previous block, and itself. The consensus method ensures that a block record is replicated as distributed ledgers, preventing tampering that may occur in many places at the same time. Each consensus mechanism method uses a computational technique to calculate the different hashes. Due to the high energy intensity of the consensus algorithms, many sectors have changed to various mechanisms such as proof-of-stake (PoS), zero-proof-knowledge (zk rollups), and proof-of-authority (PoA) in the case of electricity trading [2].

Research [4] has shown that the PoW consensus mechanism which is the most energy-intensive algorithm has the highest practice in 140 blockchain-based energy sectors. Thus, nowadays, the transformation to low energy consumption algorithms is one of the challenges and priorities in the energy sector [18, 19]. Comparison between consensus mechanisms is performed with a single-board computer in the study of Heck *et al.* [5]. It established the performance of PoW and PoA in the local energy market on the Ethereum platform which shows successful market-based solutions. PoA is also a more cost-effective alternative.

A function that accepts an entry of any size and outputs a string message of a specified size. Message integrity and digital signature systems can benefit from the usage of encrypted hash functions since they have several extra characteristics. National Institute of Standards and Technology (NIST) [6] introduced SHA family which includes SHA-0, SHA-1, SHA-2, and SHA-3. Non-linear functions are included to create complex and SHAs in which they produce output lengths of 160, 224, 256, 384, and 512 bits as shown in Figure 11.2. Within the SHA-2 family, the internal state size in SHA-256 and SHA-384 is 32 bits, while SHA-512 has 64 bits.

11.3 Simulation

11.3.1 Energy Trading Model and Case Study

Prior research by the author demonstrates the ability to trade using blockchain by SHA-256-based PoW. A total of 45 transactions had occurred including the genesis block, and the output graph of the trading can be seen in Figure 11.2 which illustrates the demand of the library building because it had demand, when the demand by energy building had not occurred. Figure 11.3

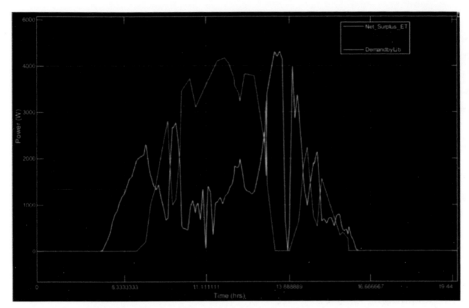

Figure 11.2 Demand power from Library building and net surplus solar power at Energy building [10].

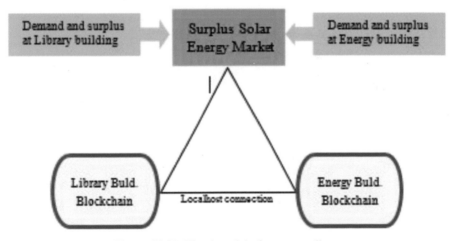

Figure 11.3 Visual model of energy trading.

depicts a visualization of the blockchain link between two prosumer models and advanced contract; equally, a smart contract is applied in the simulation.

In the model of peer-to-peer energy trading, the demand and supply energy data have been selected from one day and they are particularly used in the simulation with a real-time trading period. The peers are solar generation from the energy building which approximately had 8762 kWh and 64,550 kWh at the library building. There was sufficient surplus energy to trade between these units even after own consumption in which 1 W is equal to one token currency. MATLAB blockchain function [1] is used to execute the market model and blockchain application.

The connection between two prosumer bodies assures the reliable establishment via local host which will be used to send information and transaction data to each other. Each advanced contract will follow the protocol and energy limit and then the token which is transferred into the wallet of two parties. The surplus energy will be transferred directly to the demand side instantly after the token is deposited into the seller's wallet. The advanced contract complies with the transaction termination protocol when the buyer reaches insufficient balance and then sends information to both parties as regards the closure condition.

The hardware used in the single computer to perform the simulation is equipped with a 64-bit-based processor (Intel Core i7-10510U CPU @ 1.80 GHz to 2.30 GHz), approximately 7.9 GB of Intel UHD GPU, 9.9 GB of NVIDIA GeForce MX330 GPU, and 15.8 GB random access memory to handle the cache. The setup is used to simulate the analysis and local host is set up for peer-to-peer connection. Altering of SHAs inside the blockchain functions can be completed as *System.Security.Cryptography.SHA256/384/512Managed.*

11.3.2 Performance Result and Evaluation of the Models at Different Hash Algorithms

Two MATLAB software were used separately to perform as blockchain applications for both prosumer units. Each block has a single transaction with a total of 44 transactions being executed. An example for the block information which is data 1347 from different SHAs is applied to compare the length of the output, as seen in Table 11.1, which includes distinct nonce values, time stamps, and the hash function in hexadecimal characters.

After the successful establishment of the local connection between two blockchain platforms, performances such as usages of GPU 1 (NVIDIA GeForce MX330), GPU 2 (Intel UHD), CPU, the memory usage of GPUs, and CPU power consumption are combined and monitored during

Table 11.1 Block information of SHA-256, 384, and 512.

	SHA-256	SHA-384	SHA-512
Index	4		
Timestamp	18-August-2021 16:00:51	18-August-2021 18:01:54	18-August-2021 23:45:51
Nonce	183,732	23,452	40,524
Data	1347		
Previous Hash in Hex	0x03DCF2C4DE2086B35 3455BF07784BD528A1A 1F8A3C14CDB19BB8581 A93C6	0xD4EEB25F60B833480 486EFBF2B417D3D641A D60D52733305DD1648 4211091 DCB81A20816210FEC44 EC5B1C5BD55051F5	0x1B23B65F542EDF1B7 BD680591586C667F477 BDD82449EC4B01A747 5FB5FAAB 6414816520E76AC784A 9522C57EAE41E6244E6 F96B65075BE6DA0AB85 4E370EE11
Hash in Hex	0xF221CF3C46BD6B4EB BE36AD33EA7FC7E9792 05E77898BC0351F9F71 A05B4	0xBB1E3E4B2580A4DB1 6C0AE62C2423D14E4EF 5FA924DBF84AB061335 01D8CD3B88 394F9E7F59EFD0B094C CD4EE608EF7A	0x4AE8FB686EB663746 977CE23FBC5D2C7A4D BCFBD9750A172FECC9C CE0620A1AC 269D258CB46C85A290F 571EF3C5FFF2B1BEF97 45081D30C5AE8EA232B C178EF

transactions of two blockchain platforms every second. Heatmap correlation is used to analyze their correspondence with the transaction as visualized in Figure 11.4. The transactions in SHA-256 and 512 are positively affected by the CPU than the GPU; nonetheless, SHA-384 is not highly dependent on both but slightly by CPU.

The implementation of blockchain-based peer-to-peer energy trading based on previous historical datasets confirms the successful transactions using SHA-256, 384, and 512. The visualization is illustrated for every transaction without an idle period during the trading. As mentioned in Figure 11.5 the average transaction time for SHA 256, 384, and 512 are 71, 81, and 67 seconds, respectively. Hence, there are 3300–3500-second periods in the comparison, which is a vast data point to evaluate. The comparison is distinguished into three groups such as CPU usage, GPU usage, and GPU memory consumption. CPU usage and its power consumption have directly concerned each other and the CPU usages for three SHAs do not fluctuate where they mostly remain between 10% and 20% as in Figure 11.6. During the observation of two separate GPUs, GPU 2 has been utilized better than GPU 1 as seen in Figure 11.7. GPU 1 merely alters a small amount of period, and it does not occupy usage than GPU 2 which has a constant operation.

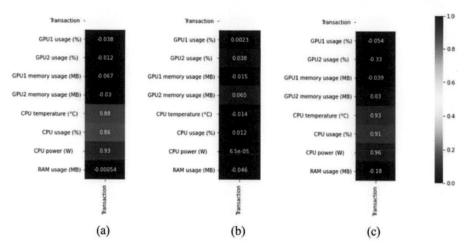

Figure 11.4 Heatmap visualization between transaction and hardware performance. (a) SHA-256. (b) SHA-384. (c) SHA-512.

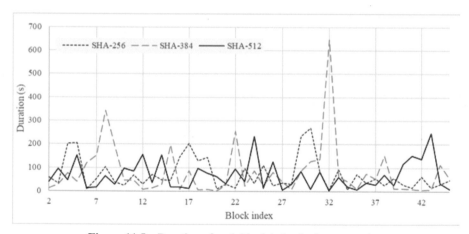

Figure 11.5 Duration of each block index in the transaction.

According to Figure 11.8, the trendline model is applied at the output to minimize the standard deviations of the dataset and simplify the performance variation between SHAs. It indicates that the CPU usage gradually decreases with time, and the consumption by SHA-512 is roughly 13% which is less than the others. CPU operation reaches 13.7% in SHA2-56 and 13.4% in SHA-384.

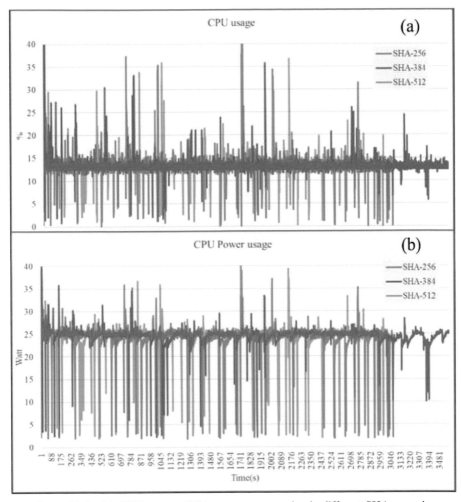

Figure 11.6 (a) CPU usage and (b) power consumption in different SHA scenarios.

As it is mentioned earlier, operation by GPU 1 has not been impacted by blockchain transactions, while GPU 2 has. GPU 1 usage is less than 0.3% for variation of SHAs; hence, it can be neglected. The linear model of GPU 2 usage indicates the noticeable changes for each SHA, where SHA-256 has operated below 0.5%, SHA-384 has the highest usage between 1.6% and 2%, and the usage is below 1.5% in SHA-512.

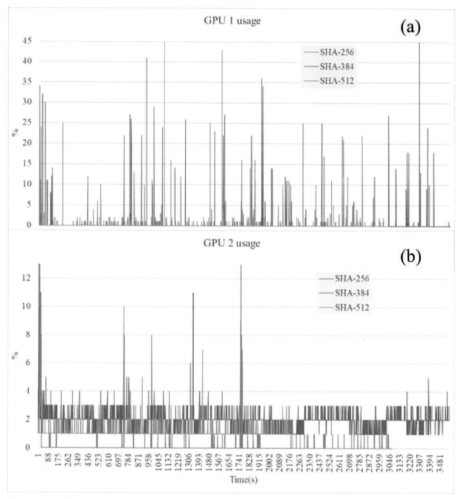

Figure 11.7 GPU usage in different SHAs. (a) NVIDIA GeForce MX330. (b) Intel UHD.

11.4 Conclusion and Recommendations

This chapter focuses on the impact of changing SHA in consensus algorithms on the hardware performance and the transaction in the blockchain-based peer-to-peer energy trading; therefore, the implementation of variation in SHA has been successfully demonstrated. Generally, the blockchain system has integrated with the SHA-256 function, but the security of the block depends on the variation of complex consensus algorithm approaches such

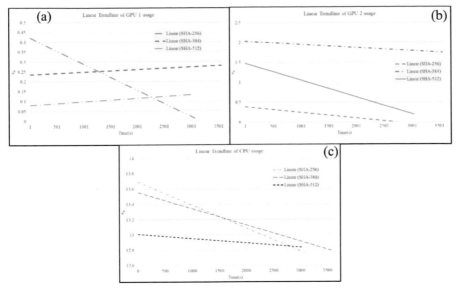

Figure 11.8 Trendline models of hardware performance in SHAs. (a) GPU 1. (b) GPU 2. (c) CPU.

as PoS, PoA, and so on. It has been pointed out that changing SHAs in blockchain-based trading impacts the GPU, CPU, and RAM performance significantly. Based on the result, the transaction system has been dominated by the CPU and RAM, which never reaches the maximum in any hashing algorithm scenario. Though they are two separate GPUs, GPU 2 performed constantly, while GPU 1 simply oscillates as a spark at some certain period. Other literature studies had measured the performance of blockchain trans- actions and some by comparing different consensus mechanisms but never in the same way as this research's approach. This research indicates the specific analysis of SHA roles in the trading model. As the nature of SHA, 512 bit is more secured than the others in the family of SHA-2 and has the longest output ASCII code. When we compare each SHA model, it cannot be indeed stated which one has the strongest and better performance. However, in a conclusion, SHA-512 has accomplished the block transactions at the lowest CPU and GPU usage at less duration than the SHA-256 and 384. Furthermore, this approach is encouraged to perform in Ethereum's platform with improved long-term energy trading system whether other findings in the aspect of security can be expected.

References

[1] Aarenstrup, R. (2021). MATLAB Blockchain Example, MATLAB Central File Exchange. Retrieved: August 23, 2021. https://www.mathwork s.com/matlabcentral/fileexchange/65419-matlab-blockchain-example

[2] Hartnett, S., Henly, C., Hesse, E., Hildebrandt, T., Jentzch, C., Krämer, K., ... & Trbovich, A. (2018). The Energy Web Chain-Accelerating the Energy Transition with an Open-Source, Decentralized Blockchain Platform. Energy Web Foundation.

[3] Nakamoto, S. (2008). Bitcoin: A peer-to-peer electronic cash system. *Decentralized Business Review*, 21260.

[4] Andoni, M., Robu, V., Flynn, D., Abram, S., Geach, D., Jenkins, D., ... & Peacock, A. (2019). Blockchain technology in the energy sector: A systematic review of challenges and opportunities. *Renewable and Sustainable Energy Reviews*, *100*, 143–174.

[5] Heck, K., Mengelkamp, E., & Weinhardt, C. (2020). Blockchain-based local energy markets: Decentralized trading on single-board computers. *Energy Systems*, *12*(3), 603–618. https://doi.org/10.1007/s12667-020-0 0399-4

[6] National Institute of Standards and Technology. (2015). *Secure Hash Standard (SHS)*. (U.S. Department of Commerce, Washington, DC), Federal Information Processing Standards Publication (FIPS), 180–184. https://doi.org/10.6028/NIST.FIPS.180-4

[7] Soshinskaya, M., Crijns-Graus, W. H., Guerrero, J. M., & Vasquez, J. C. (2014). Microgrids: Experiences, barriers and success factors. *Renewable and Sustainable Energy Reviews*, *40*, 659–672.

[8] Tao, L., Schwaegerl, C., Narayanan, S., & Zhang, J. H. (2011, May). From laboratory microgrid to real markets—Challenges and opportunities. In *8th International Conference on Power Electronics-ECCE Asia* (pp. 264–271). IEEE.

[9] Thu, K. S., & Ongsakul, W. (2020). Simulation of blockchain based power trading with solar power prediction in prosumer consortium model. In *2020 International Conference and Utility Exhibition on Energy, Environment and Climate Change (ICUE)*. https://doi.org/10 .1109/icue49301.2020.9307086

[10] Thu, K. S., Aung, M. S., Ongsakul, W., & Manjiparambil, N. M. (2020). The first national conference on engineering research. In *Transition of University to Prosumer Consortium Energy Model* (pp. 168–172). Yangon; Federation of Myanmar Engineering Societies.

[11] Santofimia-Romero, M. J., del Toro-García, X., & López-López, J. C. (2011). Artificial intelligence techniques for smart grid applications. *Green ICT: Trends and Challenges*, 41-44.

[12] Dagdougui, H., Dessaint, L., Gagnon, G., & Al-Haddad, K. (2016). Modeling and optimal operation of a university campus microgrid. In *2016 IEEE Power and Energy Society General Meeting (PESGM)* (pp. 1–5).

[13] Oh, S., Kim, M., Park, Y., Roh, G., & Lee, C. (2017). Implementation of blockchain based energy trading system. *Asia Pacific Journal of Innovation and Entrepreneurship*, *11*(3), 322–334.

[14] Plaza, C., Gil, J., de Chezelles, F., & Strang, K. A. (2018). Distributed solar self consumption and blockchain solar energy exchanges on the public grid within an energy community. In *2018 IEEE International Conference on Environment and Electrical Engineering and 2018 IEEE Industrial and Commercial Power Systems Europe (EEEIC/I&CPS Europe)* (pp. 1–4).

[15] Dib, O., Brousmiche, K.-L., Durand, A., Thea, E., & Hamida, E. B. (2018). Consortium blockchains: Overview, applications and challenges. *International Journal on Advances in Telecommunications*, *11*(1&2).

[16] Marwala, T., & Xing, B. (2018). Blockchain and artificial intelligence. *arXiv preprint arXiv:1802.04451*.

[17] Penard, W., & van Werkhoven, T. (2008). On the secure hash algorithm family. *Cryptography in Context*, 1–18.

[18] Maharjan, P. S. (2018). *Performance Analysis of Blockchain Platforms* (Doctoral dissertation), University of Nevada, Las Vegas.

[19] Zheng, P., Zheng, Z., Luo, X., Chen, X., & Liu, X. (2018, May). A detailed and real-time performance monitoring framework for blockchain systems. In *2018 IEEE/ACM 40th International Conference on Software Engineering: Software Engineering in Practice Track (ICSE-SEIP)* (pp. 134–143). IEEE.

[20] Gligoroski, D., & Knapskog, S. J. (2007). Turbo SHA-2. *IACR Cryptol. ePrint Arch.*, *2007*, 403.

[21] Hawkes, P., Paddon, M., & Rose, G. G. (2004). On corrective patterns for the SHA-2 family. *IACR Cryptol. ePrint Arch.*, *2004*, 207.

[22] Saingre, D., Ledoux, T., & Menaud, J. M. (2020, November). BCT-Mark: A framework for benchmarking blockchain technologies. In *2020 IEEE/ACS 17th International Conference on Computer Systems and Applications (AICCSA)* (pp. 1–8). IEEE.

[23] Li, Z., Kang, J., Yu, R., Ye, D., Deng, Q., & Zhang, Y. (2017). Consortium blockchain for secure energy trading in industrial Internet of Things. *IEEE Transactions on Industrial Informatics, 14*(8), 3690–3700.

[24] Hatziargyriou, N. (2014). *Microgrids: Architectures and Control.* Hoboken, NJ: John Wiley & Sons.

[25] Basak, P., Chowdhury, S., nee Dey, S. H., & Chowdhury, S. (2012). A literature review on integration of distributed energy resources in the perspective of control, protection and stability of microgrid. *Renewable and Sustainable Energy Reviews, 16*(8), 5545–5556.

Index

About the Editors

P. Sivaraman holds a B.E. in Electrical and Electronics Engineering and an M.E. in Power Systems Engineering from Anna University, Chennai, India in 2012 and 2014, respectively. He has more than seven years of industrial experience in the field of power system analysis, renewable energy, power quality and harmonic assessments, and microgrids, providing techno-economical solutions to various power quality problems for industries all over India. He is an expert in power system simulation software like ETAP, PSCAD, DIGSILENT POWER FACTORY, PSSE, and MATLAB. He is a working group member of various IEEE standards and task forces. He is a senior member of the Institute of Electrical and Electronics Engineers (IEEE), a member of the International Council on Large Electric Systems (CIGRE), and an Associate Member of the Institution of Engineers (India). He received Professional Engineers (PEng India) certification from the Institution of Engineers (India).

C. Sharmeela holds a B.E. in Electrical and Electronics Engineering, M.E. in Power Systems Engineering from Annamalai University, Chidambaram, India, and a Ph.D in Electrical Engineering from Anna University, Chennai, India in 1999, 2000, and 2009, respectively. At present, she holds the post of Associate Professor in the Department of EEE, CEG campus, Anna University, Chennai, India. She has 20 years of teaching experience and has taught various subjects to undergraduate and postgraduate students. She has done a number of research projects and consultancy work in renewable energy, power quality and design of PQ compensators for various industries. She is a senior member of IEEE, Life member of CBIP, Fellow of the Institution of Engineers (India), ISTE, and Life member of SSI, India.

Meera K. Joseph (Senior Member IEEE and Senior Member SAIEE, PMI-ITPSA) currently works as an Independent Contractor at the Independent Institute of Education, South Africa. She has been the doctoral supervisor for DBA students at the Milpark Business School from January 2021. For a short period, she served as the HoD, School of IT and Business at AIE, South Africa up to the end of December 2021. She earlier worked as the Associate Professor at the DFC, Department of Electrical and Electronic Engineering Technology, School of Electrical Engineering, University of Johannesburg (UJ) until 2018 and held various positions at University of Johannesburg, South Africa for 18 years (permanent roles). She received the degrees of D.Phil. Engineering Management (rural women and information and communication technology field), UJ in April 2014, a master's degree in Computer Applications in 1998 from Bangalore University and a B.Sc. in Chemistry (Physics and Mathematics sub.) from the University of Kerala, India. She has authored or contributed to 75 research works and has around 20 years lecturing experience in the computer engineering/ICT field. At UJ she was lecturing subjects with labs related to Java/UML, C, Javascript and HTML, Linux, VB, VBA/ VB.Net, MS Access/ SQL, C++, MS Office Applications, MS VISIO. Her multidisciplinary research interests are ICT4D (information and communication technology for development), smart grids, cloud computing, AI and machine learning for development, computer networks, femtocells, ICT for renewable energy research, ICT for power engineering research, ICT for empowerment, data analytics, Blockchain in cybersecurity and the use of ICT in engineering education.

Sanjeevikumar Padmanaban (Member'12, Senior Member'15 IEEE) received a bachelor's degree from the University of Madras, India, in 2002, a master's degree (Hons.) from Pondicherry University, India, in 2006, and a Ph.D. degree from University of Bologna, Italy, in 2012. He was an associate with various institutions like VIT University India, National Institute of Technology, India, Qatar University, Qatar, Dublin Institute of Technology, Ireland, University of Johannesburg, South Africa. Currently he is working as a Faculty Member with Aarhus University, Denmark. He is a fellow of the Institution of Engineers, FIE, India, fellow of the Institution of Telecommunication and Electronics Engineers, FIETE, India and fellow of the Institution of Engineering and Technology, IET, UK. He serves as an editor/associate editor/editorial board member of refereed journals, in particular, the IEEE Systems Journal, the IEEE Access Journal, the IET Power Electronics, Journal of Power Electronics, Korea, and the subject editor of the subject Editor of IET Renewable Power Generation, the subject editor of IET Generation, Transmission and Distribution, and the subject editor of FACTS journal, Canada.